Socially Successful Projects in the built environment

For built environment professionals, delivering socially impactful projects is more crucial and challenging than ever. This step-by-step guide provides an innovative approach to stakeholder engagement, empowering project teams to create meaningful social value through the stakeholder engagement process and build better projects.

Socially Successful Projects in the built environment will help the reader understand the theory behind a reframed approach to stakeholder engagement using a social value lens. The authors argue that projects in the built environment need to be technically feasible, financially viable, and socially successful. Project managers need to focus on all three elements, which are summarised as 'concrete, cash, and connection'. Failing to meet that third criterion can lead to pushback, protests, delays, and potentially unsuccessful projects. This book focuses on the third of those elements (the social dimension) and provides a step-by-step guide to how a project can be socially successful. In summary, this is about delivering better projects for both the stakeholders and the project teams.

In *Socially Successful Projects in the built environment*, the authors unravel the complex relationship between social value and stakeholder engagement and reveal a transformative practical theory for effective collaboration and facilitation. Along the way, they address key issues such as design philosophy, scenario development, stakeholder analysis, optioneering, and project learning, and provide practical advice on how to use these methods within real-world projects. This book is essential reading for all built environment stakeholders and teams working to deliver better projects.

Sarah Fitton is the Founder and Director of Aurora Engagements Ltd, a built environment consultancy. Sarah has over 17 years of experience working in industry in a large built environment consultancy before setting up her own firm. She also has over 13 years of experience working in academia within the fields of social value and stakeholder engagement. She graduated from Cambridge University in 2015 with a PhD examining the social value of infrastructure and continues to conduct research in this field, with five peer-reviewed journal articles and a white paper published to date and one currently under review. She is a founding member of 'Patn*' – a Manchester-based community dedicated to shaping modern culture through placemaking – and a member of the Environment Agency's Stakeholder Engagement and Communications Community of Practice (SECCOP).

Simon Wilson is the Director and co-Founder of Wilson Sherriff, a consultancy providing stakeholder engagement and facilitation expertise to clients in a range of sectors, including the built environment. Simon has over 20 years of experience working in industry with a range of clients and projects. Since 2010 much of his work has focused on projects with an environmental focus such as flood defences and water treatment sites. Simon holds an MBA degree from the Open University and is an accredited Certified Professional Facilitator (Master). He chairs the Member Council of the Organisation for Responsible Businesses.

Social Value in the Built Environment
Series Editors: Ani Raiden, Martin Loosemore and Suhair Alkilani

Social Value in Construction
Ani Raiden, Martin Loosemore, Christopher Gorse and Andrew King

Social Value in Practice
Ani Raiden and Andrew King

Constructing a Consumer-Focused Industry
Cracks, Cladding and Crisis in the Residential Construction Sector
David Oswald and Trivess Moore

Socially Successful Projects in the built environment
Demystifying social value and stakeholder engagement
Sarah Fitton and Simon Wilson

Socially Successful Projects in the built environment

Demystifying social value and stakeholder engagement

Sarah Fitton and Simon Wilson

Routledge
Taylor & Francis Group
LONDON AND NEW YORK

Designed cover image: Shutterstock

First published 2026
by Routledge
4 Park Square, Milton Park, Abingdon, Oxon OX14 4RN

and by Routledge
605 Third Avenue, New York, NY 10158

Routledge is an imprint of the Taylor & Francis Group, an informa business

British Library Cataloguing-in-Publication Data
A catalogue record for this book is available from the British Library

ISBN: 978-1-041-11416-1 (hbk)
ISBN: 978-1-041-11195-5 (pbk)
ISBN: 978-1-003-65987-7 (ebk)

DOI: 10.1201/9781003659877

Typeset in Times New Roman
by codeMantra

Contents

Figures

Tables

Foreword

Climate change is one of the greatest global challenges we face today, bringing extreme weather with more floods and droughts. To reduce the pace of change and mitigate the impacts on livelihoods and infrastructure, we draw on incredible technical expertise; engineers, spatial planners, landscape architects, scientists… all who contribute to amazing solutions.

Yet…

There remain delays and challenges to the design and construction of projects. This is not due to the complexity of the problem, but often from overlooking the social dimensions of a project.

In 2009 I managed a project which drew together the best evidence available on why and how professionals should work collaboratively with stakeholders to reduce flood risk and coastal change. This work was born from many project issues and preventable costs linked to poor engagement. Since then, I have carried out more research, created training and guidance, but still see projects start without good conversations with stakeholders. I still hear doubts and hesitation about collaboration.

So, why does this happen when we have such brilliant project managers and technical experts?

Working with local stakeholders, developing those essential relationships does take time. It can be unpredictable. It can expose issues beyond the remit of projects. It can result in disagreement and the need for compromise. But it results in the best outcomes. For everyone.

In this book, Simon and Sarah have drawn together decades of experience and knowledge to show why collaboration and engagement work. They acknowledge that it can be difficult, though not as costly as the alternative. They appreciate that it takes practice and confidence, so provide the ideas and opportunities for project managers and technical experts. Most importantly, they recognise that stakeholders' values can affect how they respond to projects but show how to uncover and understand those values to great effect.

This book demonstrates the societal benefits of engaging through ten clear principles. Their first principle states 'Built environment projects have a long lifespan affecting this and future generations. Take time to understand what matters to people who live, work and play in a place'. I defy anyone to read this book and remain unconvinced that it's worth building relationships and seeking out the needs and thoughts of others.

Dr Jacqui Cotton
Head of Flood Risk Research, Environment Agency

Preface

The aim of this book series is to present a sequence of multidisciplinary books, from a range of perspectives, that address the many ways in which 'social value' can be created (and potentially destroyed) in and by the built environment. We heartily welcome this new book on 'Socially Successful Projects in the built environment'.

The authors define a socially successful project as one where the people who are meant to benefit from a project feel that they have done so and have had a voice in the process. As the authors point out, the social aspects of projects are too often neglected in favour of technical and financial considerations which are generally seen as more pressing, easier to manage and measure and report on.

The aim of this book is to address this imbalance in management practice by bringing a practical lens to the subject, drawing on numerous real-life case studies and examples. These are based on the vast experience of the authors who have both worked within the built environment industry most of their careers. The authors have also drawn on examples from other sectors and international cases, emphasising the global relevance of socially successful projects.

In this book, the authors uniquely merge the closely related concepts of social value and stakeholder engagement to demonstrate how these concepts can be used to create socially successful projects. In doing so, they not only address a gap in the literature but also contribute new practical insights to the creation of a built environment industry that supports resilient communities, mitigates climate change, and advances progress towards the Sustainable Development Goals (SDGs). Their work is of value to industry leaders, policymakers, and community organisations seeking tools for engagement, as well as academics and students across project management, architecture, planning, engineering, and sustainability.

This book is split into four sections and fourteen chapters. Beginning with a fictional project (Chapter 1), it highlights the consequences of ignoring social factors. It then defines the built environment (Chapter 2), explaining its economic, social, and inter-generational importance, and rising community expectations for involvement.

Social value, an emerging concept linked to sustainability, the UN SDGs, and the UK Social Value Act (2012), is unpacked (Chapter 3) alongside related terms like ESG and social licence to operate. Stakeholder engagement, its undervaluation, and barriers to success are explored (Chapter 4), followed by calls for a new approach prioritising social alongside technical and financial aspects (Chapter 5).

This book introduces a socially centred design philosophy (Chapter 6) that values emotions, trust, and community perspectives. Practical guidance on early, inclusive engagement, active listening, and psychological safety is offered (Chapter 7), along with tools like situational analysis, horizon scanning, and PESTLE for understanding project contexts (Chapter 8).

Stakeholder mapping is reimagined using a social value lens (Chapter 9), while optioneering guidance (Chapter 10) shows how to embed social perspectives into decision-making. Engagement should continue throughout the project lifecycle (Chapter 11), with monitoring, evaluation, and learning ensuring ongoing improvement and evidence for social value (Chapter 12).

Future implications of AI for engagement and data use are discussed (Chapter 13), stressing the need to balance technology with human trust. The final chapter (Chapter 14) reworks the original fictional project to illustrate the book's methods, concluding with a call to create projects that are technically sound, financially viable, and socially successful.

This book arrives at a critical moment, when governments, investors, industry practitioners, and communities are demanding not just infrastructure and buildings, but outcomes that deliver long-term value for people and planet. It is an excellent addition to the Routledge series on Social Value in the Built environment, and a timely call to action. It invites practitioners, policymakers, and scholars to rethink how we design, deliver, and evaluate projects so that the built environment truly becomes a foundation for resilient, equitable, and sustainable communities.

We hope you enjoy reading this excellent and important addition to the Routledge Book series on social value in the built environment.

The Series Editors

Ani Raiden, Martin Loosemore, and Suhair Alkilani

Acknowledgments

We have been talking about writing this book for a few years now, but while working on a project together, we went out for a curry one night and actually committed to writing it. Our shared passion for putting people first on projects and championing the social dimension has led us to work together frequently. We have been lucky to work with some brilliant clients, some of whom we talk about in the following chapters, who have been willing to change their tried and tested processes and adopt new methods. We have spoken to so many people in both industry and academia who understand the importance of the social dimension, but don't know where to begin in practice. So, we decided it was time to write down our thoughts to help others do the same.

One of the most interesting aspects of writing this book has been the conversations we have had with such a diverse and interesting set of people working in this area. We were keen for this book to be not just our own thought piece, but to reflect different opinions and perspectives among practitioners and academics. Therefore, before beginning to write, we embarked on a number of conversations that helped shape the contents and direction for this book. A big thank you goes to those who provided such rich, insightful, and enjoyable conversations: Mike Adams, Arcadis, UK; Ismail Ben-Achour, Egis, France; George Berry, Berry Architects, Canada; Professor Sara Bice, Australian National University, Australia; Dr Cath Brooks, Environment Agency, UK; Emily Clarke, Binnies, UK; Dr Jacqui Cotton, Environment Agency, UK; Phillippa Dart, UK; Jo Dobson, Useful Projects, UK; Professor Toby Dubois-Heys, School of Digital Arts, UK; Julie Foley OBE, Environment Agency, UK; Salome Hoog-Chanut, Egis, France; Dr Kirsty Jones, Australian National University, Australia; Will McBain, Arup, UK; Dr Cara Mulholland, Manchester Metropolitan University, UK; Kirsty O'Connell, Australian National University, Australia; Professor Edmund Penning-Rowsell OBE, Middlesex University and University of Oxford, UK; Dr Karen Potter, Open University, UK; Dr Catherine Queen, University of Liverpool, UK; Karen Saunders, Icarus Collective, UK; Will Shaw, Binnies, UK; Camille Simon, Conservatoire du Littoral, France; Martin Stockley, Engineer, UK; Chloe Suttle, North Norfolk Council, UK; Nikky Wilson, Environment Agency, UK; Katie Wray, Deloitte, UK.

A big thank you must also go to our critical friends: Dr Jacqui Cotton, Dr Maria-Christina Georgiadou, Dr Cara Mulholland, Dr Karen Potter, and Martin Stockley. They gave up their time reviewing the first draft of this book and provided some very insightful and challenging comments which helped strengthen the final outcome, for which we are very grateful.

We also must thank each other as co-authors. When we embarked on writing we made a pact to not put pressure on each other, to enjoy this process and to air any issues before they grew into something that could derail this book. The process hasn't been easy with childcare and caring responsibilities, running two businesses and life in general. However, it has been such a positive and enjoyable experience writing with each other, testing ideas, building narratives, and seeing how this book has developed and come to life.

A final thank you goes to our families and friends who have provided unwavering support when we have either talked nonstop about social value, stakeholder engagement, and the built environment (and they weren't sure what we were talking about!) or we have disappeared for hours and days on end, locked away writing. This book is dedicated to you.

SF/SW

Introduction

> If we cannot define what we mean by value, we cannot be sure to produce it, nor to share it fairly, nor to sustain economic growth.
>
> (Mazzucato, 2018 p.Xix)

I.1 What is a successful project?

We believe that a successful project needs to be:

- Technically feasible: it needs to solve the problem that has prompted the creation of the project.
- Financially viable: the project team solves the technical problem within the financial resources they have.
- Socially successful: the project leaves a positive legacy so that it meets the needs and improves the lives of stakeholders and communities (see Figure I.1).

All three aspects should have equal status for a project to be classed as truly sustainable and therefore help achieve the project outcomes (Purvis, Mao, and Robinson, 2019).

However, the social aspect of a project is often neglected (Bostrom, 2012). Technical and financial considerations are seen as more pressing, easier to manage

Figure I.1 The three aspects of a successful project.

Source: Authors' own work.

DOI: 10.1201/9781003659877-1

and measure, and report on. This has pushed the social aspect to the back of the queue because it is harder to measure and report on (Raiden and King, 2022).

This matters for projects in the built environment because their ultimate objective is to meet human needs.

Road networks in Roman times served the social purposes of transport, trade, and social connection (English Heritage, 2025). They also had a lasting influence on transport networks. Areas that have prominent Roman road and rail networks today generally have lower construction costs for new roads or maintenance of existing ones (Dalgaard et al., 2022).

In the hot London summer of 1858, the smell from the River Thames was so pungent due to the sewage that businesses, including Parliament, were disrupted. There was an epidemic of cholera from which thousands died. This led to Sir Joseph Bazalgette masterminding the London sewer system that provided, and still provides, sanitation and health benefits to the people of London (Historic England, 2019).

Today's built environment projects meet societal needs: housing, workplaces, hospitals, transport infrastructure, flood risk management infrastructure, and energy systems. But typically, the technical and financial aspects are seen as more important than the social aspect (Purvis, Mao, and Robinson, 2019).

I.2 What is a *socially* successful project?

Put simply, a socially successful project is where the people who you intend to benefit from the project feel that they benefit. The result of the project is a structure that does its job and is appreciated, valued – even loved – by the people who use it, pass by it, or view it from their window. During the lifetime of the project the people affected by it are involved and have an influence on the final results.

Socially successful projects save you time, money, and hassle. They give you a warm feeling inside, because they are appreciated and bring a better quality of life to the people who need them. And, we would argue, you end up with better projects as you benefit from the knowledge, insight, and understanding of a wider group of stakeholders than just your project team (however expert they may be) (Denicol, Davies, and Krystallis, 2020).

We have come across many projects that are not socially successful. Problems include (Bal et al., 2013; Bice, Neely, and Einfeld, 2019):

- Stakeholder opposition.
- Additional costs.
- Programme delays.
- Resentment by the stakeholders.
- A negative legacy.
- Reputational damage.

An academic systematic literature review study by Denicol, Davies, and Krystallis (2020) examined 86 journal articles to understand the main causes of failure on megaprojects. Five main causes were identified, one of which was engaging with stakeholders and communities impacted by the project. Projects often failed due to a lack of engagement, communication, and transparency. This generated negative feelings from the stakeholders and communities leading to opposition to the project (Denicol, Davies, and Krystallis, 2020).

We want to avoid these negative effects and help project teams and individuals create socially successful projects which contribute to sustainable development.

I.3 Why socially successful projects matter

We think that this really matters. In the grand scheme of things, the built environment can help tackle the great challenges humanity faces and, if we get things wrong, can make things worse.

The built environment sector is responsible for at least 37% of global greenhouse gas emissions (United Nations Environment Program, 2023). The IPCC (2023) advocates sustainable development to help mitigate climate change impacts, reduce greenhouse gas emissions, and adapt to the changing climate. This covers the whole project life cycle, including initial planning, retrofit and reuse, and the legacy for stakeholders and communities.

Creating a sustainable built environment requires stakeholders to work together in a fair, just, and inclusive manner, taking on board diverse views and different types of knowledge. This can result in 'locally appropriate and **socially** acceptable solutions' (IPCC, 2023, p.32).

The IPCC (2023) argues that sustainable, and climate resilient, development can help to achieve the Sustainable Development Goals (SDGs). These are

> an urgent call for action by all countries – developed and developing – in a global partnership. They recognise that ending poverty and other deprivations must go hand-in-hand with strategies that improve health and education, reduce inequality, and spur economic growth – all while tackling climate change and working to preserve our oceans and forests.
>
> (United Nations, 2025, para.1)

The built environment is central to achieving most of the 17 SDGs. For example, it can help contribute to achieving the following:

- SDG 11 – Sustainable Cities and Communities emphasises the need for inclusive and net-zero urban development.
- SDG 9 – Industry, Innovation, and Infrastructure promotes the design and construction of resilient infrastructure.
- SDG 10 – Reduced Inequalities within and among countries.
- SDG 13 – Climate Action to tackle and mitigate climate change impacts.
- SDG 17 – Partnerships for the Goals highlights the importance of partnership working, stakeholder inclusion, and cross-sector collaboration to achieve fair and sustainable built environments (Georgiadou and Fitton, *under review*).

The big question though is: How do we design and construct a sustainable built environment, if we aren't doing this already? How do we create projects that mitigate climate change and achieve the SDGs? We live in the UK where the recent 10 Year Infrastructure Strategy (HM Treasury, 2025) pledged £725 billion of funding for infrastructure development over the following decade. How can we ensure that the money will be well spent?

We also believe that socially successful projects matter at the local and indeed individual level. The built environment can be a source of joy or despair … even anger. Involving people in the design and building of projects where they live makes a joyful outcome more likely.

I.3.1 Main ingredients of socially successful projects

This book looks at the two main ingredients needed to create socially successful projects:

- The first is a concept, a way of thinking: this is social value.
- The second is a method, a way of doing things: this is stakeholder engagement.

Social value is the idea that your project will benefit from knowing what matters to people – what they value – and thinking about how your project can take account of those values. If you're planning to build a bridge over a stream, what do people value about that place? Is it the convenience and ease of access? Is it the aesthetics? Is it the history of the place and the stories people tell about it? Understanding what matters to people contributes to your thinking about design, construction, and maintenance.

Stakeholder engagement is the way you can find out what matters to people and include these social values in your project at all stages, from planning, design, procurement, and construction, through to use and experience.

Additionally, understanding stakeholders' social values helps build deeper, more effective stakeholder engagement. Effective engagement helps surface and address the diverse range of social values held by different stakeholders. The relationship is complex but significant.

Understanding these complex and multifaceted concepts enables us to create socially successful projects.

I.4 Purpose, structure, and parameters of this book

We have both worked within the built environment most of our careers and the majority of our consultancy work is on built environment projects as stakeholder engagement, facilitation, and social value experts. Our experience is extensive within the water sector working with communities and practitioners on projects concerning the impacts of climate change and flood risk mitigation, hence some reliance on real-world examples from this sector within this book. We've also worked on projects relating to water quality, energy including nuclear power, broader strategies to respond to climate change, canals, weirs, pumping stations, regeneration, and other projects. We have also drawn on examples from other sectors including international cases. We believe that the principles we have drawn from our experience apply to other built environment projects.

In this book, we unpack the concepts of social value and stakeholder engagement to explore their principles, merits, and challenges, as well as the relationship between them. We demonstrate how these concepts can be used to create projects that are not only technically and financially successful, but also socially beneficial. Through this work, we hope to contribute, however modestly, to the creation of a built environment that supports resilient communities, mitigates climate change, and advances progress towards the SDGs.

This book is for people who work on projects in the built environment, whatever discipline or profession you may badge yourself as. We hope that it will also be useful for students and academics in the fields of built environment, architecture, environment, and business. We have tried to base the content on reputable research and live practice. We have tried to bridge the gap between theory and practice and offer advice on good practice.

This book is split into four sections. The first section (Chapters 1–4) sets out the theory behind social value and stakeholder engagement, examining the principles, merits, and challenges of using these two concepts on projects in the built environment. We offer our perspective on how well these approaches are used on projects today and assess the gap between theory and practice.

In the second section (Chapters 5 and 6), we propose our new approach to built environment design and construction and explain why this new approach can help to achieve socially successful projects.

In the third section (Chapters 7–10), we provide an overview of practical tools, methodologies, and guidance on how to turn theory into practice. We provide a step-by-step guide featuring different tools and approaches.

We finish this book in the fourth section (Chapters 11–14) by looking to the future at how we can create a socially successful project throughout the lifetime of a project as well as the importance of monitoring and evaluation and learning for future projects. We also look at how Artificial Intelligence (AI) can influence the social aspect of built environment projects.

We have peppered case studies and examples throughout this book to demonstrate that this approach isn't a 'nice to have', but one that actually works with real results.

We have over 30 years of combined experience working on built environment projects and in academia. We've felt firsthand the disconnect when theory hasn't translated into practice, and we've seen projects go wrong. Too often, we've experienced the frustration of local communities and stakeholders who feel ignored, who feel a project is being done to them, *not* with them.

We've also been disheartened when project teams overlook valuable data gathered from stakeholders, when technical priorities override the social dimension, or when no budget is allocated for meaningful stakeholder engagement.

And we've seen it done well, with great results.

This is what prompted us to write this book. We've seen what works, and what doesn't … and it's time to change. There is a model that can be applied to any built environment project. It requires a mindset shift and a challenge to current practice, but when done well it can save time, money, and frustration while leading to better outcomes.

This book isn't intended to be a comprehensive history of stakeholder engagement or social value in the built environment. Rather, it aims to offer practical guidance and introduce a new way of thinking for your project. You can read it cover to cover, or dip into individual chapters based on the specific challenges you are facing.

We have benefited from some great insights from colleagues while writing this book. One stuck in our minds, from Martin Stockley, a highly experienced engineer with many years of experience working on high-profile projects, who commented on the importance of the social dimension of projects:

> All of the infrastructure projects I have worked on have in some way fallen foul of this point [the social dimension]. Crossrail, Thames Tideway Tunnel, HS2 etc have brilliant technical processes at the top of their class. The social value of the processes however has been generally poor to mediocre. We haven't started conversations in the right place, in the right time, with the right clarity of aims etc. and this has diminished them all as public infrastructure projects. They are all highly successful and sustainable once complete but have been socially negative in almost all ways in their process of making.
> (Stockley, 2025)

We want to help spark a change in how the built environment is conceived and delivered. We want to put people back at the heart of projects rather than numbers, codes, or standards. We want to help leave behind a built environment that is fit for purpose for future generations. We believe that '[d]rawing on diverse knowledges and cultural values, meaningful participation and inclusive engagement processes – including Indigenous knowledge, local knowledge, and scientific knowledge – facilitates climate resilient development, builds capacity and allows locally appropriate and **socially acceptable solutions**' (IPCC, 2023, p.32).

We hope you join the socially successful project movement.

References

Bal, M., Bryde, D., Fearon, D. and Ochieng, E. (2013) Stakeholder engagement: Achieving sustainability in the construction sector. *Sustainability* 6, pp. 695–710. https://doi.org/10.3390/su5020695

Bice, S., Neely, K. and Einfeld, C. (2019) Next generation engagement: Setting a research agenda for community engagement in Australia's infrastructure sector. *Australian Journal of Public Administration* 78:2, pp. 290–310. https://doi.org/10.1111/1467-8500.12381

Bostrom, M. (2012) A Missing Pillar? Challenges in theorizing and practicing social sustainability. *Sustainability Science Practice and Policy* 8:1, pp. 3–14. https://doi.org/10.1080/15487733.2012.11908080

Dalgaard, C.-J., Kaarsen, N., Olsson, O. and Selaya, P. (2022) Roman roads to prosperity: Persistence and non-persistence of public infrastructure. *Journal of Comparative Economics* 50:4, pp. 906–916.

Denicol, J., Davies, A. and Krystallis, I. (2020) What are the causes and cures of poor megaproject performance? *A Systematic Literature Review and Research Agenda Project Management Journa*l 51:3, pp. 328–345. https://doi.org/10.1177/8756972819896113

English Heritage *Romans: Networks* [Online] Available at: https://www.english-heritage.org.uk/learn/story-of-england/romans/networks/. Accessed 02/07/2025.

Georgiadou, M. C. and Fitton, S. L. (under review) Delivering the UN Sustainable Development Goals through Inclusive Stakeholder Engagement: The Case of Flood Alleviation Schemes in England.

Historic England (2019) *The story of London's sewer system* [Online]. Available at: https://heritagecalling.com/2019/03/28/the-story-of-londons-sewer-system/. Accessed 02/07/2025.

HM Treasury (2025) *UK Infrastructure: A 10 Year Strategy.* HM Treasury: London. https://assets.publishing.service.gov.uk/media/6853c5db99b009dcdcb73649/UK_Infrastructure_A_10_Year_Strategy_Web_Accessible.pdf

IPCC (2023) Summary for policymakers. In: H. Lee & Romero, J. (eds.) *Climate Change 2023: Synthesis Report. Contribution of Working Groups I, II and III to the Sixth Assessment Report of the Intergovernmental Panel on Climate Change [Core Writing Team].* IPCC: Geneva, pp. 1–34. https://doi.org/10.59327/IPCC/AR6-9789291691647.001

Mazzucato, M. (2018) *The Value of Everything: Making and Taking in the Global Economy.* Allen Lane: London.

Purvis, B., Mao, Y. and Robinson, D. (2019) Three pillars of sustainability: In search of conceptual origins. *Sustainability Science* 14, pp. 681–695. https://doi.org/10.1007/s11625-018-0627-5

Raiden, A. and King, A. (2022) *Social Value in Practice.* Routledge: Oxford. https://doi.org/10.1201/9781003024910

Stockley, M. (2025) *Correspondence with the authors.* Unpublished, 18/8/2025.

United Nations (2025) *The 17 Goals.* [Online] Available at https://sdgs.un.org/goals. Accessed 26/06/2025.

United Nations Environment Programme (2023). *Building Materials and the Climate: Constructing a New Future.*

Section 1

Setting the scene

1 Sound familiar?

1.1 Ambridge flood risk project

1.1.1 Year 1

1.1.1.1 July

Gemma felt quietly satisfied as she reviewed the project plan she was presenting at the Project Board meeting.

The team was coming together nicely, and she had drawn upon some great data analysis and technical modelling for an innovative proposal to tackle the flood risk in Ambridge. Critically, the modelling incorporating new climate change scenarios pointed to a worsening position which strengthened the case for action and made funding more likely. Building on work by her finance adviser she had included a table of funding options that looked positive.

Initial conversations with the planning authority had been encouraging and with a fair wind they might be able to break ground within a year, enabling the initial work to be done before the weather turned.

1.1.1.2 September

The Project Board meeting had gone well. One of the members had suggested presenting the plans in a public meeting. With many members of the team away on holiday in August it had been a bit of a rush, but Gemma had booked a room in a local hotel and was now finalising the slide pack. Richard from the technical consultancy had just provided some more detailed data on long-term risk, and the business case summary was robust. She was looking forward to sharing their innovative thinking with the local community.

1.1.1.3 October

At the Project Board meeting there was some concern about the public meeting. Gemma and Richard were feeling quite bruised by the reaction. Many of the issues raised seemed irrelevant. There were a lot of grumbles about the lack of notice, the

DOI: 10.1201/9781003659877-3

timing of the meeting during working hours, and even the hotel which people said was difficult to reach by public transport. The residents didn't seem to grasp the importance of the climate change projections and quite a few were even saying that there was no real problem with flooding. A local parish councillor had spoken quite angrily and at length about local heritage and clearly didn't understand the proposals. The Project Board agreed that Gemma and Richard should attend the upcoming meeting of the parish council to put their fears to rest as time was marching on, there wasn't much wiggle room in the programme schedule, and everyone wanted to get moving with the project.

1.1.1.4 November

Lynn, the Chair of the Project Board, had called an urgent review meeting following the parish council meeting. They were all very gloomy about the headline in the local paper which described how local residents had tried to gatecrash the parish council meeting as they had heard that a secret meeting was taking place to try to railroad through the flood scheme. A spokesperson for a newly formed 'local action group' was quoted in the paper, saying that the plan to build 'a six-foot-high concrete wall in our treasured park' was unnecessary as no floods had happened there for years. Following the meeting, the Chair of the parish council had phoned the Chair of the Project Board to say that the parish council could not support the scheme in its current form. Gemma was despondent, fed up with being shouted at, and feeling that her professional judgement counted for nothing. Richard was providing more and more data to prove that their initial judgement was right. At the end of the meeting Lynn said that it would be difficult to move forward with so much opposition. Another public meeting was the only option given the rumours about secrecy. This time they needed to do some 'myth busting' to expose the misinformation that was flying around. Gemma was charged with organising the meeting.

1.1.2 Year 2

1.1.2.1 February

Given the holiday break, the earliest option for the public meeting had been January. This time they held the meeting in the village hall which was easier to access. It was a foul cold evening and Gemma, Richard, and even Lynn had been very nervous about what they might face. They believed that they had marshalled the arguments well but felt on the back foot from the off. The very first thing that happened was that the Chair of the parish council asked to read out a statement of the council's opposition to the scheme. He was applauded by members of the local community. The 'myth busting' presentation seemed to have no effect, and one after the other local community members described the negative effects of the 'concrete wall' on their park, how it would ruin the views, and destroy a favoured spot for their children to play. Gemma tried to explain that the scheme would achieve biodiversity net gain and bring in more wildlife but was accused of being 'more interested in butterflies than people'. The message just wasn't getting across.

1.1.2.2 April

As Chair of the Project Board, Lynn felt that they couldn't simply drive through a project that had become politically toxic, so she suggested a change of tack. They would meet the organisers of the action group and try to win them over. It took some time to organise the meeting which involved Lynn, Gemma, and Richard. The action group organiser was Bob, a retired businessperson, and he brought along Lianne, another local resident, a retired structural engineer who had her own ideas for how to tackle the potential flood risk. This would involve burying a pipe underground and pumping floodwater away. Richard explained that the consensus these days was in favour of 'passive' schemes and he and Lianne had a lengthy argument about the details of the scheme. No agreement was reached. Later that day they discovered that Lianne had recorded the meeting and had posted the recording on the group's Facebook page along with a biased summary of what had occurred.

1.1.2.3 July

Gemma was in a meeting with Lydia, her boss. Lydia had brought the team together for a stock take. She was a good boss and was not allocating any blame, but she asked Gemma for an honest assessment of where they were with the project. Gemma was pleased to get it all off her chest. She shared her frustration:

'We're trying to do something positive, to meet a major environmental challenge, but at every step the community is trying to stop us. They've got the parish council on their side, the media on their side, and our name is mud in the community. We can push ahead despite the opposition but I'm worried that people will lie down in front of the bulldozers. And personally, I feel that I've spent a year of my life banging my head against a brick wall. We are way over time, and the budget was blown a long time ago'.

Lydia said:

'We have no option but to pause this project. It's no one's fault but it's not working. We're going to have to go back to the drawing board and take a new approach. We need to work on the basis that this project is only going to work if it's technically feasible, financially viable – and also socially successful'.

'Socially successful?' said Gemma 'what does that mean?'

None of the details of this story are real, but all the events have happened in projects we have been involved in. Some of this may be familiar to you. Many projects in the built environment begin with positive intentions to do good in the world, create a strong technical plan and financial business case … and run into trouble when they start to engage stakeholders and communities. This is what we call the social dimension of projects.

Projects neglect the social dimension at their peril. The delays and reputational damage Gemma experienced are all too common. What is said less often is that getting the social dimension right, early in the project lifecycle, can improve not just the smooth running of projects but also achieve better results.

That's what we mean by socially successful projects. Let's first look at why projects in the built environment matter so much.

2 The built environment and why it matters

We shape our buildings; thereafter, they shape us.
(attributed to Winston Churchill, 1943)

2.1 What is the built environment?

The built environment impacts all our lives daily. In this chapter, we look at five characteristics of the built environment:

- Meeting social needs.
- Economic prosperity.
- Longevity of projects.
- The intra- and inter-generational nature of built environment projects.
- Legacy.

First, it's important to clarify what we mean by the term built environment. Our aim is not to provide a definitive definition or contribute to the academic debate, but simply to help provide context for the theory, processes, advice, and guidance that follows.

The *Cambridge Dictionary* (2025) describes the built environment as 'the parts of the places in which we live that have been built by people, for example, buildings and streets, rather than the parts that exist in nature'.

Hassler and Kohler (2014, p.120) suggest 'The built environment includes man-made building and infrastructure stocks that constitute the physical, natural, economic, social and cultural capital'.

Seyedrezaei et al. (2023, p.1) expand further to define the built environment as including 'human-made buildings and infrastructure that provide physical settings for human activities. It plays an essential role in addressing the basic needs of society, such as having places to live, work, learn, travel, and entertain'.

So, in summary the built environment refers to the manmade environment in which members of society conduct all aspects of their lives. For the purposes of this book, we see the built environment encompassing our homes, workplaces, cities and villages, transport and energy networks, other built infrastructure, recreational and leisure spaces, and more.

DOI: 10.1201/9781003659877-4

There is an interface with the natural environment. Where one begins and the other ends is not always clear. For example, the natural features of a river tributary may cause flooding in a village. There are many options for tackling that risk. A concrete defence is clearly part of the built environment. Natural flood management measures might include 'leaky dams' made of wood. These are still part of the built environment, created by human agency to change the environment where people live. Another approach to tackling flood risk might be to plant trees to delay water runoff and increase the amount of water entering the soil. Does that stand of trees become part of the built environment? You can argue it either way. However, as Seyedrezaei et al. (2023) assert, the core focus is the places we live, work, and play.

2.1.1 ***Built environment projects***

According to the Association for Project Management (APM) (2025), 'a project is a unique, transient endeavour, undertaken to achieve planned objectives, which could be defined in terms of outputs, outcomes or benefits'.

The classic distinction between 'projects' and 'business as usual' states that projects are time limited, while business as usual is continuous. If we take the example of a wind turbine generating power which is transmitted to our homes via the grid, the day-to-day operation of the turbine is business as usual. Upgrading the turbine to become more effective would be a project. Again, there are blurred boundaries. If the process of upgrading a group of turbines is built into regular operations, is it business as usual or a project? This will depend on how it is managed.

The APM also put forward three key elements or 'building blocks' for all projects: time, cost, and quality (APM, 2025). The time 'building block' derives from the time-limited nature of projects and is often presented as a Gantt chart or project plan showing when work will be performed. The cost 'building block' relates to how the project is financed and how budgets are managed. The quality 'building block' considers how to assure both the project deliverables and how the project is managed.

Thinking about sectors can help understand what constitutes a built environment project.

It is safe to say that if you are working on a project that impacts the environment a person is living in, working in or playing in then, chances are, you are working on a built environment project. The built environment includes the infrastructure, construction and building (residential, commercial, retail, mixed use, etc.) sectors, and also:

- Water including flood risk, waste management, and water treatment.
- Transport including rail, roads, aviation, cycling, and walking.
- Healthcare.
- Energy infrastructure including renewables.
- Digital services including data centres.

This list is not exhaustive. In this book, we draw in particular on our experience in the water sector including flood risk, waste management, and water treatment as well as other examples and experience from different sectors.

2.2 Five important characteristics of built environment projects

As we mentioned at the start of this chapter, we have identified five characteristics which have a major impact on the success of built environment projects. These are:

- Meeting social needs.
- Economic prosperity.
- Longevity of projects.
- The intra- and inter-generational nature of built environment projects.
- Legacy.

Let's look at each one in turn.

2.2.1 Meeting societal needs

The built environment helps meet individual and societal needs. The places we live, work, socialise, and play in are important for meeting our everyday needs. They also affect our health and well-being.

Cities and urban environments provide the backdrop and setting for communities to come together and thrive. Cities that are well laid out become destinations for different activities: walking, meeting people, having conversations, working, and living (Gehl, 2010). They have a major economic impact as hubs for education, business, and employment (OECD, 2024).

The role of cities, towns, and villages as social spaces is vital given the current loneliness epidemic, which the World Health Organisation (WHO) has classed as a public health emergency. The WHO (2025) compares the effects of chronic loneliness to the physical health impacts of smoking, obesity, and physical inactivity. A poll by the Centre for Social Justice (2024) found that nearly six in ten adults in the UK (nearly 32 million people) felt lonely some of the time with 22% of adults feeling 'existentially' lonely.

The built environment has the potential to help or hinder crises like these. It can be shaped to help alleviate loneliness, social isolation, inequalities, and physical inactivity. The built environment has the power to create spaces that enhance people's well-being both physically and mentally. A simple example is the Danish city of Århus where the river was filled in and converted into a street for vehicular traffic in the 1930s. Sixty years later the river was uncovered and the spaces along the reopened waterway were laid out as recreational pedestrian areas. Since then, the areas along the Århus river have been the most commonly used external space in the city (Gehl, 2010).

Opening up spaces in this way creates opportunities not just to combat loneliness but for individual and collective self-actualisation similar to what Maslow described as the highest element of a hierarchy of personal needs. Only after basic physiological and safety needs are met can people and societies fulfil their potential (Maslow, 1954). By analogy, the bottom two levels of the hierarchy – physiological needs, and safety and security – can be met by the built environment through water infrastructure, sanitation systems, homes, healthcare facilities, transport networks, access to employment opportunities, and places to work. Meeting these needs provides the foundation for people and societies to fulfil their potential.

The societal value of the built environment encompasses not only buildings and public spaces but also the infrastructure that supports everyday life. Features such as transport networks, sewage systems, and waste disposal are often overlooked, yet they are the unsung heroes that make shared spaces functional and pleasant.

While we often associate the built environment with towns and cities, these features are equally relevant in smaller settlements or recreational areas like sea bathing spots and walking trails.

Media and industry reports frequently highlight the consequences of inequality, such as poor transport links, limited access to healthcare, and inadequate educational facilities, on children's life chances and overall community well-being. This is why built environment projects must go beyond technical success. They also need to be socially successful: enhancing individual well-being and enabling people to reach their full potential.

2.2.2 Economic prosperity

The built environment, and more specifically, built environment projects, can also provide economic benefits at a local, regional, and national level. The World Green Building Council (2025) states that the creation of sustainable environments in which people can live, work, and play can help develop resilient thriving communities as well as economic growth through job creation, skill development and training opportunities, economic investment, and economic regeneration. The actual design and construction process of built environment projects also stimulates growth through job creation, investment in the supply chain, and investment opportunities (CIOB, 2025).

Shops, factories, sports grounds, and concert venues are built structures that house economic activity. 'Destination' structures like cathedrals, castles, or the bridges and canals of Venice bring people and income to places. Infrastructure is key to this. The London Underground created new living spaces and increased economic activity when it built stations in the middle of fields in the 19th and 20th centuries (Wolmar, 2020). The stations brought with them housing, shops, and work extending the economic reach of the city.

Complex environments depend on infrastructure. Amin and Thrift (2017) in their book *Seeing Like a City* explain how the areas where we spend our time are interconnected systems including everything from buildings to transport, to energy, to communications, to data processing systems. It is all vital for these places to work effectively.

In the world of flood and water management, a massive example is the River Thames flood barrier which has been estimated to protect £321 billion worth of property in Greater London (Environment Agency, 2021). The barrier also brings social benefits and, referring to Maslow's hierarchy in Section 2.2.1, perhaps most importantly a sense of safety and protection for people living and working in London.

2.2.3 Longevity of built environment projects

Most built environment projects have long lifespans, including the planning, design, and construction phases and the actual lifetime use of the project once built.

The longevity of the built environment means that we need to consider carefully the project's effects on the environment, health, resource consumption, sustainability, and climate change, to name but a few. Get a project wrong and it will probably stay wrong for decades or even centuries.

If we flip that – get a project *right* – the positive impacts felt will reach far and wide for decades or even centuries.

The example of the Basilica de la Sagrada Familia Catholic church in Barcelona highlights this. It is one of the most popular tourist attractions in the city. Five generations have experienced the planning and construction of this historic masterpiece. The project was originally designed in 1882 by Francisco de Paula del Villar with the cornerstone being laid in March of that year. However, in 1883, the famous and flamboyant architect and designer Antoni Gaudí took over and took the project in a very different direction – the standout architecture and design that is synonymous with Barcelona today. Gaudí's vision for the church meant the scale of the project was vast and therefore the construction timeframe was long. So long, in fact, that the church is still not formally completed. The APM would probably not approve! The church is open and usable and is a big tourist attraction for the city, but parts are still under construction with access restricted. The central tower of Jesus Christ is the last tower to be erected and is due for completion in 2026, meaning the project will have taken around 144 years to complete from design through to construction (Basilica de la Sagrada Familia, 2025).

Or take the High Line railway, built in 1934 to carry trains delivering goods to businesses in New York's industrial district, fulfilling a strong economic need. It also improved the safety of deliveries as street-level rail deliveries caused so many accidents early in the century that the route was known as 'death avenue' (New York City Department of Parks & Recreation, 2025). As road freight began to take over, rail traffic declined, and the line was abandoned by the 1980s. A local campaign to save the railway from demolition was led by a non-profit group, Friends of the High Line, who proposed to turn the track into a park. The High Line park opened in sections between 2009 and 2023 featuring a 'keep it wild' planting scheme and artwork commissioned and displayed every year. It has become one of New York's most popular tourist attractions, bringing a different type of economic and social benefit from the original rail tracks (Institution of Civil Engineers, 2025b).

Both examples point to the way that built environment projects may need to meet changing needs as the economy and society evolve. The Sagrada Familia began as a religious building, the High Line as a rail line. Both evolved over time to meet different needs and both became tourist attractions. This leads us to our next characteristic.

2.2.4 Inter- and intra-generational nature of the built environment

As we discussed above, some projects span decades and even centuries in planning, design and construction, and use. Often the built environment has to meet the needs and aspirations of different ages and generations.

Some projects are intra-generational as they have to serve the needs of current generations which can range from the young to the elderly (Abdulmunem et al., 2024; Kaplan and Haider, 2015). For a housing development this means meeting needs ranging from mobility to access and social inclusion. This is quite an ask of a project and also of those responsible for the planning, design, and construction.

On the other hand, some infrastructure projects are inter-generational (Kaijser, 2004). This means they will span and serve generations way into the future. Transport and energy systems are examples of projects designed, constructed, and implemented before current generations were born which will probably still be functioning after current generations have gone. This raises a really important question: How can current professionals – engineers, architects, planners, modellers – understand and plan for generations that aren't even alive yet?

Houses built in the UK in the 1920s and 1930s were the first to have garages for private cars. Many of those houses are still lived in today, but the garages are generally too small for modern cars. On the other hand, modern cars need less protection from the elements, so often garages are converted to living spaces and the cars sit outside on the drive. Could planners and developers a hundred years ago have done more to envisage these changes? More importantly, how good are we today at thinking ahead to the technological developments and social needs of the next hundred years?

In this book, we argue that we need to incorporate the social needs, values, and perspectives of communities impacted by a project. This is achievable by asking questions of and collaborating with the people who will ultimately be the end users. We must also ask: How do we talk to, question, and collaborate with multiple and future generations?

2.2.5 Legacy

All built environment projects leave a legacy. We have already seen how the Basilica de la Sagrada Familia has impacted five generations through its design, construction, and part opening, with no doubt many more in the future.

The Pruitt-Igoe housing complex in St Louis, Missouri, in the United States has a legacy for the wrong reasons. The complex of 33 eleven-storey high-rise towers was opened to fanfare between 1954 and 1956 but lasted only 20 years, with the towers demolished between 1972 and 1976. Poor design and social changes in the city led to crime, vandalism, and degradation of the building stock which became vandalised and vermin-ridden (St Louis Area Office, 1974). The project left a complex legacy. It is sometimes seen as a case study of the failure of modernist design, but its context was one of economic decline and racial discrimination in the housing policy and practice of the time (Bristol, 1991).

We all want to work on projects that have a legacy we can be proud of. We want our own legacy to be a built environment that enhances and positively impacts those who use it now and who will use it in the future. We don't want to be remembered for holding back the race to net zero, worsening social exclusion, or simply for being part of projects that fail.

Built environment projects should leave a legacy for generations to come that supports and provides them with the best living, working, and playing environments (as much as possible given that we can't ask them what they want). Industry professionals should be proud to put their names to planning applications, drawings, and construction contracts that will become part of the history of that project.

2.3 Expectations are growing

The built environment is also important because those who live, work, and play there say it is. Greater access to information and the ability for anyone to find reports, drawings, and planning applications on the internet mean that community voices have grown louder over the years. Knowledge is power, as they say, and access to information is the public's greatest power in these circumstances. PlaceSpeak, a Canadian website, is an example which aims to connect residents in communities to influence planning decisions with evidence-based contributions (PlaceSpeak, 2025).

People no longer accept being told what is being done – there is a growing sentiment that people want a say in what is happening to the places that affect their lives. A 2024 survey in the UK found that three in five respondents wanted more public involvement in the planning system (ISEP, 2024), and other similar surveys showed the same. The ability to request and demand information means people who aren't necessarily industry professionals can gain knowledge and formulate opinions as to whether a project is acceptable or not to them. This consequently allows them to either support the project or oppose it. People and communities are not willing to accept poorly designed projects, or projects that they perceive will have a negative impact. If feelings are strong enough, community opposition can be significant. In Chapter 6 we consider what this can mean for your project.

Key takeaways

Built environment projects impact everyone: this and future generations. They take time to plan, design, and construct and may well be around for decades, even centuries in some instances.

Buildings and structures affect people and communities in different ways, and those effects change over time.

People increasingly want a say in what happens to the places that matter to them. Therefore, it's important to understand, as far as possible, the needs and values of the people who live, work, and play in a place.

References

Abdulmunem, A. S., Shok, M. E., Hussain Al-Hussaini, Z. I. and Alkinani, A. S. (2024) Multigenerational urban design: Creating urban spaces that support active aging and intergenerational interaction. *International Journal of Sustainable Development and Planning* 19:10, pp. 3737–3747. https://doi.org/10.18280/ijsdp.191004

Amin, A. and Thrift, N. (2017) *Seeing Like a City.* Polity Press: Cambridge. https://doi.org/10.1111/1468-2427.12618

Association for Project Management (2025) *What is project management?* [Online] Available at: https://www.apm.org.uk/resources/what-is-project-management/. Accessed 07/05/2025.

Basilica de la Sagrada Familia (2025) *History of the Basilica* [Online] Available at: https://sagradafamilia.org/antoni-gaudi. Accessed 30/4/2025.

Bristol, K. G. (1991) The Pruitt-Igoe Myth. *Journal of Architectural Education* 44:3, pp. 163–171. https://doi.org/10.1080/10464883.1991.11102687

Cambridge Dictionary (2025) *Built environment* [Online] Available at: https://dictionary.cambridge.org/dictionary/english/built-environment. Accessed 01/05/2025.

CIOB (2025) *Socio-economic impact of construction* [Online] Available at: https://www.ciob.org/industry/policy-research/policy-positions/socio-economic-impact-construction. Accessed 1/5/2025.

Environment Agency (2021) *Thames Estuary 2100: 10-year monitoring review (2021)* [Online] Available at: https://www.gov.uk/government/publications/thames-estuary-2100-te2100-monitoring-reviews/thames-estuary-2100-10-year-monitoring-review-2021. Accessed 08/05/2025.

Gehl, J. (2010) *Cities for People.* Island Press: Washington.

Hassler, U. and Kohler, N. (2014) Resilience in the built environment. *Building Research and Information* 42:2, pp. 119–129. https://doi.org/10.1080/09613218.2014.873593

Institute of Sustainability and Environmental Professionals (ISEP) (2024) *UK public wants more involvement in planning process, ISEP research finds* [Online] Available at: https://www.isepglobal.org/articles/uk-public-wants-more-involvement-in-planning-process-iema-research-finds. Accessed 14/8/2025.

Institution of Civil Engineers (2025b) *High Line, New York* [Online] Available at: https://www.ice.org.uk/what-is-civil-engineering/infrastructure-projects/high-line-new-york. Accessed 14/9/2025.

Kaijser, A. (2004) *The Dynamics of Infrasystems. Lessons from History.* 6th International Summer Academy on Technology Studies, Graz, Austria.

Kaplan, M. and Haider, J. (2015) *Creating Intergenerational Spaces that Promote Health and Well-being in Intergenerational Space*. Routledge: London. https://doi.org/10.4324/9780203736920-3

Maslow, A. H. (1954). *Motivation and Personality*. Harper and Row: New York. https://doi.org/10.4236/psych.2025.161003

New York City Department of Parks & Recreation (2025) *The High Line* [Online] Available at https://www.nycgovparks.org/parks/the-high-line/history. Accessed 14/8/2025.

OECD (2024) *Regions and cities at a glance* [Online]. Available at: https://www.oecd.org/en/publications/oecd-regions-and-cities-at-a-glance-2024_f42db3bf-en/full-report.html. Accessed 23/08/2025.

PlaceSpeak (2025) *Welcome to placeSpeak: Citizen engagement you can trust* [Online]. Available at: https://www.placespeak.com/en/. Accessed 14/8/2025.

Seyedrezaei, M., Becerik-Gerber, B., Awada, M., Contreras, S. and Boeing, G. (2023) Equity in the built environment: A systematic review. *Building and Environment* 245, pp. 1–19. https://doi.org/10.1016/j.buildenv.2023.110827

St. Louis Area Office (1974) *Pruitt Homes and Igoe Apartments Public Housing Complex: Final Environmental Impact Statement.* St. Louis Area Office: Kansas City. https://doi.org/10.7936/K7F18X0T

The Centre for Social Justice (2024) *Lonely Nation - Part 3: How to tackle loneliness through the built environment.*

Wolmar, C. (2020) *The Subterranean Railways: How the London Underground was Built and How it Changed the City Forever.* Atlantic Books: London.

World Health Orgainsation (2025) *Social isolation and loneliness* [Online] Available at https://www.who.int/teams/social-determinants-of-health/demographic-change-and-healthy-ageing/social-isolation-and-loneliness. Accessed 01/05/2025.

World Green Building Council (2025) *What is a sustainable built environment?* [Online] Available at: https://worldgbc.org/what-is-a-sustainable-built-environment/. Accessed 01/05/2025.

3 Social value

The emerging buzzword

> Cities have the capability of providing something for everybody, only because, and only when, they are created by everybody.
>
> (Jacobs, 1993 p.238)

3.1 Where did social value come from and what is it?

In this chapter, we begin to unpack the concept of social value, a widely used term across the built environment, but one that is often bandied about without a clear understanding of its meaning.

Social value is rooted in the concept of sustainability. Sustainability was also a buzzword for the built environment 30 or 40 years ago. Today sustainability is ingrained in everyone's lives from professional work and the legislation that governs it, to personal lives, travel choices, and consumer habits. For a project to be truly sustainable, there needs to be equal consideration of the three elements of sustainability: social, environmental, and economic. Social value is critical in understanding the social element of sustainability. By understanding the theoretical background, how it can be defined and the different ideas associated with the term, we hope to provide a bit more clarity to a complex concept.

The theory of social value is rooted in different disciplines including psychology, philosophy, sociology, and geography, all providing their own perspectives (Graham et al., 2013; Ives and Kendal, 2014). It's both interesting and complex, and still evolving.

The process of social value measurement is not dealt with in depth in this book. We do touch on it in Chapter 12 regarding measuring and monitoring, but it is only a 'lite' touch. However, it should be acknowledged that there are tools available to measure social value and it is important to understand what you want to measure in order to choose the most appropriate approach.

3.2 Defining social value

Although social value has gained prominence over the years within the built environment, the concept itself is still ambiguous and confusing (Raiden et al., 2019). There is no one agreed term to describe social value within the built environment.

DOI: 10.1201/9781003659877-5

As a result, varying definitions have been created for different contexts. The lack of clarity concerning its meaning, and the theoretical premise on which to base definitions, is often a barrier to its implementation (Fitton, 2015; Raiden et al., 2019).

There is no consensus in industry guidance on how social value should be articulated (Raiden et al., 2019). Project teams working in the built environment have a choice of definitions. Let's look at a few.

'It [social value] moves beyond using money as the main indicator of value, instead putting the emphasis on engaging people to understand the impact of decisions on their lives' (Institute for Social Value, 2025).

'Social Value is created when buildings, places and infrastructure support and enhance the quality of life of the people who live, work, play and learn there' (Stuart, 2024).

'[Social value is] … the holistic social, environmental, and economic benefits to society that a project creates' (University of Reading and RIBA, 2020, p.3).

> Social Value is all about the positive impact an organisation has on society beyond its financial bottom line. While profit is tracked in traditional accounting terms, social value is measured by the actions an organisation takes to improve the world around it.
>
> (Social Value Portal, 2024a)

The definitions all have similarities:

- They are people focused.
- Social value is about doing better and bringing change to people and society.
- Social value is about more than just economic and environmental considerations.

However, there are subtle differences. Some definitions say it's about enhancing what is already there, some say it is about positive impacts created, others focus on public spending.

There is no right or wrong answer. As there is no single industry definition, no definition can be considered entirely right or wrong. For us and our work, we use the definition by the UK Green Building Council (2021, p. 6) 'In the context of the built environment, social value is created when buildings, places and infrastructure support environmental, economic and social wellbeing, and in doing so improve the quality of life of people'. This provides the built environment frame that helps us engage with our clients and stakeholders. A big question though is: Does social value really need to be defined?

We would argue that no, it doesn't need a definition. It is more important to understand the concept of social value than to strive for a definition. Moving away from seeking a one-size-fits-all definition would increase understanding and encourage people to use the concept more meaningfully.

The *concept* of social value is about understanding the impact that a project or organisation brings to a group of stakeholders, a community or society as a whole. Each stakeholder has different beliefs and values, so the impact will be different for each of them. In the built environment we may identify desirable values including increasing well-being, job creation, improved air quality, flood protection, resilience to threats such as climate change, promotion of social inclusion, and use of local supply chains (ICE and Useful Projects, 2020; RICS, 2020; University of Reading and RIBA, 2020). Cara Mulholland, Lecturer in Project and Operations Management and a social value specialist at Manchester Metropolitan University, urges caution when trying to understand social value. She explains that social value can be perceived to do everything and cover a lot, but this doesn't need to be the case for it to still be utilised meaningfully. The high-level concept of social value needs to be fragmented so that it isn't seen as an all-encompassing silver bullet answer. It is important that the parameters for the stakeholder and the project are clear (Mulholland, 2025).

3.2.1 How did this ambiguous term land in the built environment?

Projects in the built environment often pay more attention to the economic and environmental elements and neglect the social element of sustainability. In part this is because social sustainability is the least defined and least understood of the different elements of sustainability (Fitton and Moncaster, 2022). Social sustainability has had considerably less attention in public dialogue, and academic research, than economic and environmental sustainability. So, we tend to see technically successful projects with degrees of economic and environmental sustainability, but few technically *and* socially successful projects.

Consequently, over the last few years, attention has turned to addressing social sustainability and trying to rebalance the three elements. One way to do that is through the concept of social value.

3.3 The theory of social value

So, what is the theory behind social value? This is the £1 million-pound question. The answer, quite honestly, is that it is still a relatively young concept being addressed from different schools of thought. There is little published research on social value, but the body of knowledge is growing slowly. In academia the theory and concepts are not being disseminated as fast as industry is moving. There is a practice/theory gap emerging. Practitioners are making the best of their understanding of social value without strong theoretical underpinning evidence and professional training. However, the theory behind how social value is constructed allows us to piece together a picture of how it can manifest in practice.

O'Brien and Wolf (2010, p.233) suggest that the term 'value' relates to: 'interests, pleasures, likes, preferences, moral obligations, desires, wants, goals, needs, aversions and attractions'. Tsirogianni, Sammut, and Park (2014) state that social value is a standard created by groups and individuals in order to shape the social order of a collective and define what is right, what is wrong, what is acceptable and what is not. Values are inherently subjective as they are personally created by each individual (O'Brien and Wolf, 2010). A person's values are constructed through their own experiences and develop as a person grows (Rokeach, 1979; Shillito and De Marle, 1992). Our living and working environments shape our value preferences: what we like, what we don't, what we are willing to accept, what we aren't willing to accept. Therefore, because our experiences are constantly changing, our values are continually changing and evolving (O'Brien and Wolf, 2010).

Our values act as criteria on which a person bases their judgements, choices, arguments, preferences, likes, and dislikes. These ultimately create a person's evaluation criteria for what they believe is right and wrong (O'Brien and Wolf, 2010). Our perception of what is right and wrong, in turn, influences how we act in everyday situations (Ives and Kendal, 2014; Rokeach, 1979). Therefore, social value in the built environment relates to the value we see in a project.

Social value is a difficult and somewhat ambiguous concept because of how we experience things. The way in which a person experiences something is always unique and varies between individuals. People perceive experiences, even the same experiences, differently due to their experiences and knowledge, so people have their own individual values (Rokeach, 1979). People who live in the same or similar environments may have some shared experiences, so there may be shared organisational or community values (O'Brien and Wolf, 2010). But individuals will also have unique social values and may respond in different ways to their environment and experiences (Camarinha-Matos and Macedo, 2010; Rokeach, 1979). Social value is a multi-layered concept.

It seems like the right time for an example. In the small town of Bude in Southwest England, climate change and coastal erosion are threatening homes, businesses, and assets in the town. Faced with many concerns from various threats, the local community decided to focus their efforts on preserving Bude's historic storm tower from coastal erosion and subsequent destruction as there were only a few metres between the tower and the cliff edge at some points. It was built in 1835 as a refuge for the coastguard and is a Grade II listed building.

The community raised £60,000 in contributions and received grant funding to move the tower, stone by stone, 100 m inland to save it for years to come. The move was completed in 2024.

Some people questioned the local community's priorities. You might argue that there are more pressing needs around present-day housing, businesses, or community spaces. Yet in choosing this target for fundraising and preservation the community were asserting what they valued, and their choice will become part of history. They valued their cultural heritage and preserving it demonstrated its social value (Heritage Fund, 2021).

3.4 Drivers for social value

The concept of social value is being heralded as a means to achieving the United Nations Sustainable Development Goals (UN SDGs). In their book *Social Value in Practice*, Ani Raiden and Andrew King (2022) demonstrate how social value is a key practical tool to helping achieve the SDGs. The SDGs are a 'call to action to end poverty, protect the planet, and ensure that all people enjoy peace and prosperity' (Raiden and King, 2022, p.6). This is a huge task. Raiden and King (2022) discuss how by considering and creating social value, the SDGs can be achieved. However, they can't be achieved solely by government or large organisations. Everyone, including local communities and individual households, has a part to play in adapting, shifting, and changing behaviours in order to achieve change. Social value on all levels helps to create a better society.

The Public Services (Social Value) Act 2012 promoted the term 'social value' in England. The Act requires local authorities and procuring bodies to consider social factors alongside economic and environmental factors in procurement scores (Cabinet Office, 2025a). Emerging from procurement and its processes, it acted as a catalyst for the broader consideration of social value. In England, the National Procurement Policy Statement (Cabinet Office, 2025b) moved the focus of procurement away from just buying and more towards supporting local needs and developing social and economic value. The Statement is focused on procurement issues and therefore neglects the broader remit of social value within the built environment, but it is progress from previous procurement guidance which was heavily focused on the creation of jobs as a means to delivering social value.

In conjunction with the Procurement Policy Statement (Cabinet Office, 2025b), a National Procurement Note (PPN002) on the use of the UK Government's Social Value Model was released in 2025 (GOV.UK, 2025). The Social Value Model, launched in 2021, is a framework to help public sector buyers create questions and evaluate answers from suppliers as to how socially focused their practices are (Samtaler, 2025). The updated Social Value Model has a greater focus on aspects of social value such as well-being and community cohesion (Social Value Portal, 2025).

Other countries have also adopted social value-orientated policies and legislation, although England seems to be the only country explicitly developing social value legislation. Wales has the Wellbeing of Future Generations Act (2015) (Welsh Government, 2025) which seeks to improve the social, economic, environmental, and cultural well-being of Wales and the Welsh people. The Act ensures that development takes into account the legacy of projects and promotes engagement and collaboration with affected people and communities (Welsh Government, 2025). The Victorian Government in Australia has developed the Social Procurement Framework. This legislation seeks to ensure that procurement goes beyond value for money and maximises opportunities for social and sustainable outcomes for the communities and people of Victoria (Victoria State Government, 2025). In Vancouver, Canada, the City Council has a Community Benefit Agreements Policy. This policy requires developers to support the needs and growth of the local communities they are working in (City of Vancouver, 2025).

We would argue that it is important for project teams to understand the concept of social value as this will underpin the strategy or approach for embedding it in your project. The precise definition is less important. As we will see in the next section, social value has many associated concepts and for this reason we suggest that social value is actually an umbrella term under which many different strands hang.

3.5 Related concepts

How does social value align with other buzzwords floating around in the sustainability sphere? What other concepts are sheltering beneath the social value umbrella?

You may be familiar with terms such as social impact, Environmental, Social, Governance (ESG), Corporate Social Responsibility (CSR), social accounting, or even Social Licence to Operate (SLO). All these terms refer to understanding the positive impact an organisation or project will have on a group of stakeholders, a community or society. The terms are often used interchangeably with social value as they all demonstrate social value principles (Raiden et al., 2019). However, they are different, and it is useful for your project, and the project team, to understand the differences. Is social value the right concept for your project? Or do you actually mean CSR? Or is the governance aspect important and therefore ESG is a more appropriate concept?

Let us take a look at a representative sample of some of these terms. The list is by no means exhaustive. We have chosen them because they are the ones we have come across most frequently, but you will find many others.

3.5.1 Environmental, Social, Governance

The Social Value Portal (2024b) explains how the ESG agenda evolved in response to the need for investors to understand the environmental, social and governance risks of a business's operating model. ESG covers aspects such as the supply chain, engagement with stakeholders, ethical business processes and the environmental and societal risks that may affect shareholder value. Typically, a business will develop an ESG strategy to document these considerations as well as how positive impacts can be emphasised and risks mitigated (Social Value Portal, 2024b).

There is overlap between social value and ESG, not least the importance placed on stakeholders. ESG is an important concept in ensuring businesses act in a socially responsible way. However, its focus is limited compared to the broader scope of social value. ESG is risk focused, centred on the organisation. Social value has a wider focus, stretching beyond the organisation towards all those affected by the business or project. Social value is more proactive in its creation of positive social outcomes rather than a reactive response to risks like ESG.

The European Union's Corporate Sustainability Reporting Directive (CSRD) has prompted more alignment between ESG and social value. Businesses are required to report their impact on the environment and society against the European Sustainability Reporting Standards. Adopting social value principles is a way of identifying those impacts and reporting on them under the CSRD (Social Value Portal, 2024c).

Many companies today are publishing their ESG reports. A typical example is Pigeon, a real estate investment company based in the UK. Their ESG statement is perhaps quite typical. Environmental objectives include '100% renewable power by 2030'. The People/community objectives include 'wellbeing support for all employees' and the Governance objectives include 'all directors receive regular training and professional development support' (Pigeon, 2024).

3.5.2 *Corporate Social Responsibility*

CSR is probably the most well-known term due to its longevity and prevalence. This is also the concept (or model) that is used most interchangeably with social value.

CSR uses the three elements of sustainability (economy, environment, society) to capture a business model for organisations to prioritise people, the planet and profits in equal measure. The people aspect addresses the social element of sustainability. The purpose of the CSR model is to demonstrate to a variety of stakeholders – including investors, employees and consumers – the ethical considerations of the organisation (Impact, 2025).

CSR and social value are often used interchangeably because examples of CSR activities are often similar to the attributes of social value. For example:

- Local community initiatives.
- Reduction in carbon footprint.
- Local charity donation in the form of money and time.

Peel Ports Group is the second largest port operator in the UK. Some of their CSR commitments include (Peel Ports Group, 2025):

- Providing job and financial security.
- Enhancing skills via education, apprenticeships, and training.
- Driving regeneration in areas of traditional deprivation.
- Protecting and championing environmental developments.

The difference between CSR and social value lies in the focus. CSR takes a business perspective, while social value has a more outward facing (built environment, community) perspective. Social value focuses on long-term positive impacts on stakeholders, communities, and society, rather than the organisation creating them.

3.5.3 *Social impact and social accounting*

Social impact refers to the social outcomes that result from socially conscious decision-making. This can be from an organisational perspective (overlapping with CSR) or from a project perspective. Social impact differs from social value because social value consciously aims for a better impact, or more value than before. Activities are consciously targeted at improvement. Social impact is more of an assessment of what results from decision-making, even if the aim was not to create social value.

For example, the insurance company Aviva reports on impact through the independent Aviva Foundation. In 2024 it reported the following impacts (Aviva Foundation, 2024):

- Provided over £2 million in funding to organisations.
- Over 14,000 people directly received support designed to strengthen their resilience supported by the Aviva Foundation.

- Over 34,000 people indirectly received support designed to strengthen their resilience supported by the Aviva Foundation.

Social impact can also be viewed from a project perspective, as well as an organisational perspective. The International Association for Impact Assessment (IAIA) explains that a social impact assessment is a process of analysing, monitoring and managing the deliberate and unintended human consequences, both positive and negative, of a project (IAIA, 2025). These assessments are usually produced as part of the project's suite of documents during the design phase. In some cases, the social impact assessment is carried out in conjunction with the economic impact assessment in what is called a socio-economic impact assessment.

An example of this can be found in plans for the Roberts Bank Terminal 2 project, a new container terminal project in British Columbia, Canada. The project is situated in the traditional territory of several indigenous groups. WSP, a global built environment consultancy, were contracted to complete the assessment and they prioritised engaging early with the indigenous groups to understand the social impacts of the project. They defined the social, economic and health and well-being values of importance to communities as well as identifying the importance of community members accessing higher skilled and longer-term jobs. As a result of this, some members of the indigenous communities were trained in data collection and helped gather baseline data for the assessment (WSP, 2025).

Social accounting takes a more economic approach to identifying and measuring impacts. Instead of the qualitative reporting used predominantly for ESG, CSR, and social impact, social accounting focuses on turning the social impacts into monetised figures. These figures are used to show, financially, the social impacts an organisation has achieved (Torrecchia, 2013). Social value accounting is a branch of social accounting in which the qualitative social value identified and created is monetised.

The UK not-for-profit organisation Step by Step is a charity seeking to help young people who are facing difficult challenges such as homelessness. The organisation used the Social Return on Investment (SROI) model to calculate a return of £6.80 worth of social value for every pound spent on their activities. They did this by engaging with the young people to assign monetary value to factors such as the benefit of feeling physically and psychologically safer as a result of their actions and help (Step by Step, 2024).

3.5.4 Social Licence to Operate

SLO refers to the level of acceptance or the approval of a project by the local community or stakeholders that can influence the profitability of that project (Bice and Moffat, 2014). This term is prominent in Australia on infrastructure projects and especially in the mining sector. Bice and Moffat (2014) explain that an SLO can be indicative of the relationship between the communities, stakeholders and the developer as poor relationships generally do not foster SLOs. In contrast, SLOs can empower communities and stakeholders to set out their terms for accepting a project by hindering a SLO until they find the terms acceptable. Research in Peru's mining regions found that companies that did not have good relationships with the local communities found it harder to obtain a SLO. Companies that had taken the time to understand the context of the community, communicated effectively with the community and demonstrated commitment to the community found it easier to obtain a SLO (Pósleman and Sallan, 2019).

These terms and concepts are not interchangeable, although they all feature similar principles. We see them as overlapping elements all of which emphasise the social element of sustainability. It is worth considering which approach, or combination of approaches, may be most relevant to your context. If your main perspective is organisational, CSR or ESG may be the starting point, but social value and the methods we describe in this book are likely still relevant. Engaging early, listening to stakeholders, and involving them in shaping outcomes are relevant whatever your focus.

What is important is how you use these approaches to progress and enhance your project. Socially successful projects are built on understanding social value: what matters to the people, organisations, and communities affected by the project.

Key takeaways

Social value means understanding what people and communities value about a place.

There are different definitions, but the importance of social value lies in its contribution to sustainability and, in some countries, legislation.

Sometimes values are held in common, often individuals have different social values.

Related concepts include ESG, CSR, social impact, social accounting, and SLO. Social value connects them all.

Think about how the concept of social value applies to your project.

References

Aviva Foundation (2024) *The difference we're making: Impact summary*. Aviva Foundation: London.

Bice, S. and Moffat, K. (2014) Social licence to operate and social impact. *Impact Assessment and Project Appraisal* 32:4, pp. 257–262. https://doi.org/10.1080/14615517.2014.950122

Cabinet Office (2025a) *Social value act: Information and resources* [Online]. Available at: https://www.gov.uk/government/publications/social-value-act-information-and-resources/social-value-act-information-and-resources. Accessed 23/08/2025.

Cabinet Office (2025b) National Procurement Policy Statement. UK Cabinet Office: London. https://assets.publishing.service.gov.uk/media/67ab330e1a116437c7ed88da/E03274856_National_Procurement_Policy_Statement_Elay.pdf

Camarinha-Matos, L. M. and Macedo, P. (2010) A conceptual model of value systems in collaborative networks. *Journal of Intelligent Manufacturing* 21, pp. 287–299. https://doi.org/10.1007/s10845-008-0180-7

City of Vancouver *Community Benefit Agreements Policy* [Online] Available at: https://vancouver.ca/people-programs/community-benefit-agreements.aspx. Accessed 10/08/2025.

Fitton, S. L. (2015) *Social Value in Practice: A Case of Flood Alleviation Schemes*. PhD Thesis. University of Cambridge.

Fitton, S. L. and Moncaster, A. (2022) Social value, infrastructure and stakeholder engagement: A complex triangle. *Proceedings of the Institution of Civil Engineers – Engineering Sustainability* 175:4, pp. 194–201. https://doi.org/10.1680/jensu.21.00030

GOV. UK (2025) *Procurement policy note 002: The social value model* [Online] Available at: https://www.gov.uk/government/publications/ppn-002-taking-account-of-social-value-in-the-award-of-contracts/procurement-policy-note-002-the-social-value-model-html. Accessed 24/05/2025.

Graham, S., Barnett, J., Fincher, R., Hurlimann, A., Mortreux, C. and Waters, E. (2013) The social values at risk from sea-level rise. *Environmental Impact Assessment Review* 41, pp. 45–52. https://doi.org/10.1016/j.eiar.2013.02.002

Heritage Fund (2021) *Saving Bude's historic storm tower* [Online] Available at: https://www.heritagefund.org.uk/projects/saving-budes-historic-storm-tower?utm_source=chatgpt.com. Accessed 24/05/2025.

ICE and Useful Projects (2020) Maximising social value from infrastructure projects.

Impact (2025) *Cutting through the jargon: what's the difference between CSR, social impact, and social value?* [Online] Available at: https://impactreporting.co.uk/difference-between-csr-social-impact-and-social-value/. Accessed 04/05/2025.

Institute for Social Value (2025) *What is social value and why does it matter?* [Online]. Available at: https://socialvalueuk.org/what-is-social-value. Accessed 04/05/2025.

International Association for Impact Assessment (IAIA) *Social impact assessment* [Online]. Available at https://iaia.org/social-impact-assessment/social-impact-assessment-general-info/. Accessed 06/06/2025.

Ives, C. D. and Kendal, D. (2014) The role of social values in the management of ecological systems. *Journal of Environmental Management* 144, pp. 67–72. https://doi.org/10.1016/j.jenvman.2014.05.013

Jacobs, J. (1993) *The Death and Life of Great American Cities*. Vintage Books: New York.

Mulholland, C. (2025) 'Book Interview' by Sarah Fitton and Simon Wilson [Microsoft Teams], Unpublished, 11/2/2025.

O'Brien, K. L. and Wolf, J. (2010) A values-based approach to vulnerability and adaptation to climate change. *Climate Change* 1, pp. 232–242. https://doi.org/10.1002/wcc.30

Peel Ports Group (2025) *Our commitments* [Online] Available at https://www.peelports.com/sustainability/local-communities. Accessed 24/05/2025.

Pigeon (2024) *Environmental, Social and Governance (ESG) Policy Statement* [Online] Available at: https://www.pigeon.co.uk/esg-policy/. Accessed 24/05/2025.

Pósleman, C. S. and Sallan, J. M. (2019) Social license to operate in the mining industry: The case of Peru. *Impact Assessment and Project Appraisal* 37:6, pp. 480–490. https://doi.org/10.1080/14615517.2019.1585142

Raiden, A. and King, A. (2022) *Social Value in Practice*. Routledge: Oxford. https://doi.org/10.1201/9781003024910

Raiden, A., Loosemore, M., King, A. and Gorse, C. (2019) *Social Value in Construction.* Routledge: Oxford. https://doi.org/10.1201/9781315100807

RICS (2020) Measuring social value in infrastructure projects: insights from the public sector.

Rokeach, M. (1979) *Understanding Human Values: Individual and Societal*. Collier Macmillan Publishers: London.

Samtaler (2025) *Social Value Model: What is it and how to score full marks in bid responses* [Online] Available at: https://www.samtaler.co.uk/social-value-hub/what-is-social-value-model-asking-for. Accessed 24/05/2025.

Shillito, M. L. and De Marle, D. J. (1992) *Value: Its Measurement, Design and Management.* John Wiley & Sons, Inc: New York.

Social Value Portal. (2024a) *What is Social Value? Definitions and examples for businesses* [Online] Available at: https://www.socialvalueportal.com/news-and-insights/what-is-social-value-definitions-examples-and-how-to-create-it#what-is-social-value. Accessed 04/05/2025.

Social Value Portal. (2024b) *Social Value and ESG: Understanding the difference* [Online] Available at: https://www.socialvalueportal.com/business-posts/social-value-and-esg-whats-the-difference. Accessed 04/05/2025.

Social Value Portal. (2024c) *What CSRD means for businesses – and for Social Value* [Online] Available at: https://www.socialvalueportal.com/business-posts/what-csrd-means-for-businesses-and-for-social-value. Accessed 04/05/2025.

Social Value Portal (2025) *A new National Procurement Policy Statement: What does it mean for Social Value?* [Online] Available at: https://www.socialvalueportal.com/news-and-insights/a-new-national-procurement-policy-statement-what-does-it-mean-for-social-value. Accessed 24/05/2025.

Step by Step (2024) Social Return on Investment 2024.

Stuart, K. (2024) *Blog: Why is social value so crucial when developing and managing commercial real estate?* [Online] Available at: https://ukgbc.org/news/why-is-social-value-so-crucial-when-developing-and-managing-commercial-real-estate/#:~:text=What%20is%20social%20value%3F,improve%20people's%20quality%20of%20life. Accessed 04/05/2025.

Torrecchia, P. (2013). Social accounting. In: S. O. Idowu, Capaldi, N., Zu, L. & Gupta, A. D. (eds.) *Encyclopaedia of Corporate Social Responsibility.* Springer: Berlin. https://doi.org/10.1007/978-3-642-28036-8_191

Tsirogianni, S., Sammut, G. and Park, E. (2014) Social values and good living. In: A. C. Michalos. (eds.) *Encyclopaedia of Quality of Life and Well-Being Research.* Springer: Dordrecht, pp. 6187–6190. https://doi.org/10.1007/978-94-007-0753-5_3666

UKGBC (2021) *Framework for defining social value: A framework for defining and delivering social value on built environment projects.* UK Green Building Council: London.

University of Reading and RIBA (2020) Social Value toolkit for Architecture.

Victoria State Government (2025) *Introduction to the social procurement framework* [Online] Available at: https://www.buyingfor.vic.gov.au/introduction-social-procurement-framework. Accessed 10/08/2025.

Welsh Government (2025) *Guidance document: Well-being of Future Generations (Wales) Act 2015: the essentials* [Online]. Available at: https://www.gov.wales/well-being-future-generations-act-essentials-html#60668. Accessed 10/08/2025.

WSP (2025) *Socio-economic impact assessment for a proposed container terminal project* [Online] Available at: https://www.wsp.com/en-gb/projects/socio-economic-impact-assessment-for-a-proposed-container-terminal-project. Accessed 24/05/2025.

4 Stakeholder engagement

The (missed) opportunity

> Involving stakeholders in decision making is a key skill for effective leadership. It can help you gain buy-in, improve collaboration, and generate better solutions. However, it can also be challenging to balance different perspectives, expectations, and interests.
>
> (ACEP, 2023, p.6)

We believe that it is beneficial for projects to surface, and work with, the social value of the people who are affected by the project. Stakeholder engagement is one of the main ways to make this happen, but it has to be done well.

In this chapter we outline some of the key features of stakeholder engagement. There is no commonly agreed approach to stakeholder engagement underpinned by systematic academic research. This chapter is based on a combination of sources, experience, and professional practice. We then look at the factors that often prevent stakeholder engagement from having the best effect on projects.

In later chapters, we go into more detail about the practical methods that will surface social values for the benefit of projects, stakeholders, and communities.

4.1 Why involve stakeholders?

Much thinking about stakeholders and stakeholder engagement started with businesses. Businesses recognise that managers and leaders should think about groups other than shareholders when developing strategies.

A stakeholder is 'a group or individual who can affect or is affected by the achievement of the organisation's mission' (Freeman, 2010, p.52). A business may have internal stakeholders (owners, customers, employees, suppliers, shareholders) and external stakeholders (government, competitors, consumer advocates, environmentalists, special interest groups, media). '… if business organisations are to be successful in the current and future environment, then executives must take multiple stakeholder groups into account' (Freeman, 2010, p.52).

What applies to organisations applies to projects. It is easy to become focused on internal processes such as technical planning, budget setting, drawing up a business case, and pleasing the project leader. Yet all projects have a wider group of internal and external stakeholders who can affect or are affected by what the project is trying to achieve. Often this includes members of the public.

DOI: 10.1201/9781003659877-6

Stakeholders may be *affected by* the project in many ways. Building a flood gate, for instance, protects a community from flooding. It may help them feel more secure in their homes. It may also affect how community members can reach their favourite dog walking spot, or how customers for a local shop can reach it or how that shop can ensure it continues to trade without disruption when flooding takes place.

Stakeholders can also *affect* the project. We often think about the negative ways: businesses may object to the effect of the flood gate on their trading, or residents form protest groups to preserve their favoured routes. But they can also affect projects positively by bringing different perspectives and types of knowledge to the table. Long-term residents may know about historic patterns of flooding; business owners may be able to suggest alternative sites for the flood gate that would not disrupt business.

4.2 Engaging stakeholders

So, if stakeholders are important to a project, how do you reach them? This is where stakeholder engagement comes in.

> Today, the term "stakeholder engagement" is emerging as a means of describing a broader, more inclusive, and continuous process between a company and those potentially impacted that encompasses a range of activities and approaches and spans the entire life of a project. The change reflects broader changes in the business and financial worlds, which increasingly recognise the business and reputational risks that come from poor stakeholder relations, and place a growing emphasis on corporate social responsibility and transparency and reporting. In this context, good stakeholder relations are a prerequisite for good risk management.
>
> (International Finance Corporation, 2007, p.3)

Again, the reference here is to companies but the principles apply to built environment projects developed by public bodies or partnerships. The UK government guidance on stakeholder engagement emphasises the benefits of involving stakeholders. It sets out a broad framework based on three stages: defining objectives, identifying stakeholders, and developing a plan to engage them (Cabinet Office, 2021).

Some organisations we work with distinguish between communication (essentially one-way, providing information) and engagement (two-way, providing information and asking for a response). Let's look at the steps in engagement.

4.2.1 Defining objectives

Engagement objectives are often seen as a subset of the main project objectives. If your project aims to tackle flood risk by installing flood gates, it's important to consider the objectives for engagement. Are you looking to gain stakeholder views or buy-in, to minimise opposition, and/or to involve community members long term in the operation of the gates? Engagement objectives should then shape the remainder of the engagement plan and activities. At this stage, it also makes sense to identify risks specifically relating to engagement and begin thinking about how to evaluate engagement (Cabinet Office, 2021).

4.2.2 Stakeholder analysis and segmentation

Stakeholder analysis is the process of identifying stakeholders, finding out what matters to them, and deciding how to engage with them. The project team should produce a long list of stakeholders. This means thinking about different perspectives and interests within the broad definition of stakeholders: individuals, groups, or organisations who can affect or be affected by the project.

Stakeholder segmentation helps break down the long list of stakeholders into manageable categories in order to develop an engagement plan with activities targeted at different segments.

Stakeholder analysis and segmentation together are sometimes called 'stakeholder mapping'. We use that term in Chapter 9 where we explore the process in more detail.

4.2.3 Deciding on stakeholder influence

The key question is: How much influence do stakeholders have?

For projects in the built environment we see three main types of influence: inform, consult, and collaborate. We say more about these three 'rungs' below.

4.2.3.1 Inform

This is the apparently simple job of putting out information to stakeholders, considering key communication elements (Harrin, 2020).

- Who are you communicating with (the audience)?
- What are you communicating about (the message)?
- How are you communicating (the channel/s)?

4.2.3.2 Consult

Consulting means asking people their opinion about a proposal so that they can have an influence on what happens. In the UK planning system, there are detailed technical requirements about whom to consult, how and when (GOV.UK, 2022). Other organisations may decide to consult because it is good practice, and they want to hear what the people impacted have to say.

Either way, this is a complex task which involves:

- Being clear about what aspects of a project are open to influence and what have already been decided.

- Reaching the right stakeholders and gaining their attention.
- Managing expectations on the part of both stakeholders and the project team.
- Communicating the proposed changes and the issue being consulted on.
- Gathering views from stakeholders.
- Using views from stakeholders to shape the project.
- Deciding how to act on different and contradictory views from stakeholders.

4.2.3.3 Collaborate

Collaboration means organisations, teams, and/or communities working together to achieve what they cannot achieve alone. Collaborative engagement implies individuals, teams, stakeholders, or organisations keeping in close contact, usually meeting regularly, often through a partnership board or other group. Genuine collaboration is quite rare, and difficult to achieve successfully. It should only be considered where there is scope for stakeholders to go beyond having an influence to having a role in the project itself. Collaboration means being prepared to give up control in order to access the benefits of collaboration (Vangen, 2017).

Some commentators have identified other steps on the engagement spectrum, often quoting five: inform, consult, involve, collaborate, and empower (Last, 2019). In practice, we find this makes for less rather than more clarity so prefer to stick to three.

However, it is worth mentioning the concept of deliberative engagement. 'Deliberative processes are a type of participatory process that allow participants to consider and discuss relevant information from multiple perspectives before coming to a common conclusion' (Involve, 2025, p.1). Methods such as citizens' juries and citizens' panels have grown out of this approach, bringing together broadly representative groups from an area or community to deliberate on issues of public concern. In Chapter 8 we share an example of this approach.

4.2.4 Developing an engagement plan

Having identified the stakeholders and their different segments and worked out the level of involvement they will have, a good project team will put together an engagement plan.

Typically, this looks like any other project plan providing a statement of objectives to achieve, a timeline of activities with appropriate outcome and evaluation measures. The engagement plan should also use the information and knowledge gained from the stakeholder analysis to build a picture of your stakeholders and create a way forward with different groups. This will include information like the channels of communication to be used, the frequency of engagement envisaged, and any risks or challenges that you foresee. The plan helps the whole project team understand who the stakeholders are, how to carry out engagement, and how to achieve the objectives set. It should be reviewed by the project team on a regular basis, when stakeholders change or at key milestones in the project and updated accordingly.

The engagement plan can also be a useful document from which to develop engagement plans for individual stakeholders. This is especially useful if you are working on a large project with a large and diverse number of stakeholder groups.

4.3 Skills and competencies for stakeholder engagement

Different skills are needed for successful projects, and that includes the range of skills required to support stakeholder engagement. We know at least one organisation that has a capability framework for engagement professionals, but the body of skills and knowledge has not been codified in professional standards. We outline here our thoughts on what a competency framework should cover.

Engagement design and planning, project management: the competence to develop a plan including objective setting, stakeholder analysis, and identifying engagement activities, and to evaluate that plan.

Dedicated engagement and communication skills: the competence to communicate effectively with team members, colleagues, partners, and stakeholders including listening skills. Diplomacy, collaboration, and partnership work including dealing with conflict and challenge. Specialist skills in facilitating groups and handling conversations with the public.

Personal competencies and mindset: understanding how diversity and power dynamics affect engagement. The competence to use empathy and emotional intelligence to value all contributions, and the skills to elicit social value. The skill of reflective practice is needed and the competence to learn lessons and continue to improve practice.

We spoke to Chloe Suttle, an early career programme team member working on coastal transitions for a local authority in the UK, and asked what skills she saw as essential for engaging communities about future changes to the coastline including floods and coastal erosion. She mentioned flexibility, empathy, and

> being able to have hard conversations that no one really wants to have around what might be coming in future. Even if it's something that could happen ten years in the future, would you rather this community have ten years of preparation or tell them the year before something happens? We should focus on being able to have those conversations, be empathetic and give all the information that's needed.
>
> (Suttle, 2025)

4.4 What happens in practice?

In this chapter we have outlined key elements of stakeholder engagement and explained its importance. Yet in practice we often see stakeholder engagement deprioritised or done sketchily.

There are several contributing factors to why this happens.

4.4.1 The status of stakeholder engagement in project management

There is a vast amount written about how to deliver great projects. Depending on where you look, you may find references to stakeholder engagement. A classic model for projects has five stages or 'process groups':

1 Initiating.
2 Planning.
3 Executing.
4 Monitoring and controlling.
5 Closing.

It's generally recognised that it's important to engage stakeholders in the initiating stage so they can contribute to the definition of a project and so they know that a project is happening (Coursera, 2024). Typically, some form of stakeholder analysis is done at that stage, such as a RACI matrix which identifies which stakeholders are **r**esponsible, **a**ccountable, **c**onsulted, and/or **i**nformed (APM, 2025). This tends to focus on stakeholders within an organisation, while other segmentation models may be used for external stakeholders.

There are opportunities to involve stakeholders throughout a project. There is a growing recognition that stakeholder engagement is key to effective projects although at the time of writing this is still very much an unfinished revolution (Babaie, Locatelli and Sainati, 2023; Maddaloni et al., 2025).

The Project Management Institute signalled a shift in 2021 through its key guidance document, moving from the five 'process groups' outlined above to a broader framework of 'project delivery principles' encompassing both stakeholders and value (Bell, 2025).

Similarly, in the UK the Association for Project Management (APM)'s *Body of Knowledge*, a core text for project managers, now includes sections on stewardship and engagement and has also moved away from the step-by-step approach (APM, 2025).

The Royal Institute of British Architects (RIBA) model *Plan of Work* has little direct reference to stakeholders within a complex and detailed methodology for building design. RIBA recognises the importance of project stakeholders even though they do not have a contractual relationship with the project team so 'managing or responding to the range of opinions and requests can be challenging' on projects (RIBA, 2020 p.14). RIBA followed this by developing an 'engagement overlay' for the plan of work recognising that '[w]hen carried out successfully engagement helps create a sense of inclusion and leaves participants feeling they've added value, as well as supports more desirable, effective, and efficient project outcomes' (RIBA, 2025, p.5).

In our experience, these shifts in professional guidance are only slowly finding their way into practice. Many project teams still see stakeholder engagement as secondary to the technical and financial aspects of the project. It's often seen as nice to have, useful confirmation that a project is on track and that there is no significant opposition. Or it's seen as a compliance issue, particularly when it comes to the formal consultation requirements for planning regulations and therefore gaining permission for a project to go ahead.

4.4.2 Tokenism

In the 1960s, Sherry Arnstein described a 'ladder of participation' with the nirvana of 'citizen control' at the top (Arnstein, 1969). Arnstein's ladder included eight levels:

1 Manipulation.
2 Therapy.
3 Informing.
4 Consultation.
5 Placation.
6 Partnership.
7 Delegation.
8 Citizen control.

She divided these different types of engagement into 'nonparticipation' (levels 1 and 2), 'tokenism' (levels 3–5), and 'citizen control' (levels 6–8).

We find the ladder useful not because of the detail but because of the warning about tokenism. Even with the best intentions, stakeholders often feel that engagement is simply going through the motions and that their views will not have any impact.

For the flood gate example we have discussed in previous chapters, it may be challenging to explain what is open for discussion. It may be that the decision has already been made that flood gates are needed, and the question is *where* they should be sited. So there is engagement over the location. However, some stakeholders may still want to make the case against the gates. The engagement process needs to explain what is open to negotiation and what has been decided, while not shutting down views.

4.4.3 Partial engagement

Project teams need to listen to what stakeholders have to say and act on it. This can be counter cultural as the evidence gathered this way may be seen as subjective and may contradict technical analysis. Local knowledge is immensely powerful but is often overlooked by project teams in favour of apparently more robust, evidence-based technical knowledge (Babaei, Locatelli, and Sainati, 2023). Project teams are often unclear about how much influence stakeholders will have on their final decisions.

Then there is the question of how to gather views. Today a lot of consultation is done through online surveys which are a pretty blunt instrument. Not everybody is comfortable working online. Other approaches, like public meetings in town halls, may encourage participation by some but be off-putting for others. Family commitments, caring responsibilities, and working patterns all mean that not all people have the ability to attend in person events in the evening or during the day. Instead, a range of channels will be needed to account for different needs and preferences. Again, this is time consuming and costly but ensures that you gain the widest range of views.

4.4.4 Fear of stakeholder engagement

We also often come across some fear of stakeholder engagement. One project manager told us 'I don't want to go to some village hall to be shouted at by the public'. The fear comes from knowing that sometimes stakeholders will be opposed to what the project is aiming for or that the team has not explained what the objectives are. Community members don't like the idea of flood gates shutting off their usual route to the park, or think the gates will be ugly, a blot on the landscape. We also notice that in recent years different groups, and the public as a whole, have become more vocal and cannier about how to express their views. As we have said, social media has contributed to this. So, the project team may find themselves facing a campaign of outright opposition.

One response is the head in the sand approach. It's tempting not to engage stakeholders unless you really have to and only tell them what they really need to know. If they don't know what you're planning, they can't shout at you.

Another, and probably the one we encounter most, is to put off the evil day until very late in the project. We hear all sorts of rationale for this:

'We don't want to get people worried while we're still working out the details.'

'We don't have anything specific to tell them at this stage.'

'They don't have a role to play yet.'

And of course, engagement costs time and money, so it's always tempting to push it back until another day.

The risk with deprioritising or delaying engagement is that when stakeholders find out what you have in mind, the backlash is even sharper as people realise, they have been kept in the dark. Just as important, delaying engagement means missing out on the knowledge and insight of stakeholders, so useful options may be missed, while the team pursues what they think is technically preferable but may prove unacceptable to the community.

We spoke to a community member, Angharad, who told us how she was personally affected by poor stakeholder engagement. A planning application was submitted for 100 homes in her village in Wales, essentially doubling the size of the village. There was negligible engagement with the community.

Angharad explained how she felt community concerns were not listened to or taken on board. The proposal was perceived to be submitted very secretively by the developer. The result was a divided community due to a road acting as a barrier between the new houses and the existing village, trees removed, and increased traffic.

The stress brought by this poor engagement took its toll on some community members as they struggled to fight to be heard by the developers and the local authority. The result was that the original community members were very distrusting of the local authority and the developer and were very cynical of intentions regarding building within the village.

This example shows how the fear of engagement can actually backfire for project teams. Angharad believed that the developer thought it would be easier to just drive the planning application through without consulting the community, but in fact the level of opposition and the resultant damage to reputation were detrimental to both the developer and the local authority. In this case poor engagement was a missed opportunity to build trust, but it was also evident from our conversation that having your values disregarded can cause emotional distress.

4.4.5 Undervaluing stakeholder engagement skills

We find that often stakeholder engagement skills are less valued than technical skills such as engineering or even project management skills. This is analogous to the way that communication skills are sometimes described as 'soft skills'. In our experience stakeholder engagement expertise is brought in as an afterthought, often when things have started to go wrong, and it can be difficult for stakeholder engagement professionals to be heard, in particular if their advice contradicts that of other professionals.

In our projects we try to challenge the distinction between 'technical skills' and 'engagement skills'. Engagement requires specific technical skills and competencies as outlined above, and to undervalue them is to damage your project.

4.5 Complexity and challenges around engagement

Even when well supported and valued in projects, stakeholder engagement 'lands' in social contexts of complexity, and the results can be surprising.

In the flood and coastal erosion risk management field a research review focused on community engagement around the increasing likelihood of flooding and coastal erosion resulting from climate change. It identified challenges relevant to that context which may be relevant for other projects concerned with the built environment (Kelly and Kelly, 2019). The challenges were as follows.

4.5.1 Readiness

'Readiness' is about how far people are ready and able to participate constructively in planning and decision-making processes, especially on complex choices around climate change and major external factors. We worked on a project which the project team considered routine. It involved talking to the community about how to manage flood risk locally. The project team was surprised by how hard some community members pushed back. It became clear that many community members did not understand or accept the degree of flood risk, so they were not ready to think or talk about disruptive changes to manage a problem they did not recognise.

4.5.2 Framing information

Stakeholders may perceive or understand issues such as flood and coastal erosion risk differently depending on their experience or perspectives (and indeed their values). For instance, public authorities often use the word 'resilience' to mean how individuals and communities can be prepared for risks such as flooding. Yet we have found that often communities interpret this term as meaning they are left alone and abandoned by the authorities. Instead of implying strength, it implies vulnerability. However, in some cases, this vulnerability can lead to empowerment. Communities that accept that they will not have public funding or support may develop their own support network in terms of people, skills, and knowledge needed to adapt to the risk (Georgiadou and Loggia, 2023). This leads them to develop their own narrative and share information through their own networks.

4.5.3 Climate change, emotions, and mental health

How people respond to information about flood and coastal erosion risk involves a complex mix of emotions and can have mental health implications. This applies to people who have experienced flooding, but similar deep anxiety can apply to other types of risk. We worked with a community experiencing chronic foul smells from gases escaping a landfill site. Members of the community felt deep concern and anxiety for their health and their children's and were not convinced by factual scientific information.

4.5.4 Place attachment, culture, and identity

People's emotions – positive, negative, or mixed – about the places in which they live, or work shape their willingness to engage with proposed changes, their relationships with other local residents and/or organisations, their local knowledge, and their responses to professionals or facilitators coming in from the 'outside'. Every time we have worked on a project with a local focus, we have had to spend time breaking down barriers of trust to demonstrate that we understand what is unique about that place, and what matters to people who live locally.

4.5.5 Power and politics

It's important to recognise the political and power issues that influence relationships between authorities and communities. This may involve formal mechanisms of politics, such as a local authority's official attitude to a project changing overnight following an election. It also involves how power is experienced by stakeholders (Environment Agency, 2019). We worked on a project for a health body which involved doctors and patients working together to identify service development needs. The patients believed that the doctors could do 'whatever they wanted', having no idea of the constraints of budgets and procedures. Only after the doctors had explained the constraints on their actions could the two sides discuss priorities in a meaningful way. The patients then advocated for more resources to enable the doctors to meet their needs.

4.5.6 Scale

A sixth challenge was also identified around how people and organisations work at very different times and geographical scales. Long-term projects can feel like an unwelcome distraction from the here and now, so it is helpful to identify both shorter and longer-term benefits. It can be difficult to negotiate tensions from people working at different geographic scales. We worked on a project which was aiming to develop long-term adaptive approaches to meet the challenges of climate change by the year 2100. Participants in an early planning meeting found it difficult to envisage that timescale given the short-term pressures on resources and the need to respond to weather events. They wanted action 'in their lifetime'.

So, these are just some of the challenges. Stakeholder engagement has the potential to unlock social value in projects, but it is not a perfect process in itself. In the next chapters, we review how this might fit together and look closer at the relationship between these two concepts. In the second half of this book, we explain how to do it.

Key takeaways

Stakeholder engagement is not just a matter of process. It's not enough to plug an engagement plan into a project.

Engagement, and the people who work on it, are often undervalued and given low priority in projects.

At the same time, engagement requires grappling with complexity and multiple risks and challenges.

Engagement is about people, and their responses can be unexpected, emotional, and disruptive. They can also provide the key to success.

References

ACEP (2023) Developing more authentic consultation & engagement: A discussion paper by ACEP. Association of Consultation and Engagement Professionals.

Arnstein, S. R. (1969) A ladder of citizen participation. *Journal of the American Institute of Planners* 35:4, pp. 216–224. https://doi.org/10.1080/01944366908977225

Association for Project Management (APM) (2025) *Body of knowledge*. 8th Edition. Association for Project Management: London.

Babaei, A., Locatelli, G. and Sainati, T. (2023) Local community engagement as a practice: An investigation of local community engagement issues and their impact on transport megaprojects' social value. *International Journal of Managing Projects in Business* 16:3, pp. 448–474. https://doi.org/10.1108/IJMPB-10-2022-0224

Bell, M. (N.D) *PMBOK guide* 7th Edition vs 6th Edition [Online] Available at: https://projectmanagementacademy.net/resources/blog/pmbok-guide-7th-edition-vs-6th-edition/. Accessed 19/05/2025.

Cabinet Office (2021) *Ensuring effective stakeholder engagement* [Online] Available at: https://gcs.civilservice.gov.uk/publications/ensuring-effective-stakeholder-engagement/. Accessed 16/05/2025.

Coursera (2024) *Project management life cycle: The 5 phases explained* [Online] Available at: https://www.coursera.org/articles/project-management-lifecycle. Accessed 16/05/2025.

Environment Agency (2019) *Working together to adapt to a changing climate: flood and coast Key learning from an evidence review on community engagement on climate adaptation* Slide pack accessed online: https://assets.publishing.service.gov.uk/media/60916d8f8fa8f51b98b1513d/Community_engagement_on_climate_adaptation___slide_pack__3_.pdf

Freeman, R. E. (2010) *Strategic Management: A Stakeholder Approach.* Cambridge University Press: Cambridge. https://doi.org/10.1017/CBO9781139192675

Georgiadou, M. C. and Loggia, C. (2023) Community-led vs. subsidised housing. Lessons from informal settlements in Durban. *Housing Studies* 40:11, pp. 1–28. https://doi.org/10.1080/02673037.2024.2397425

GOV. UK (2022) *Guidance Consultation and pre-decision matters* [Online] Available at: https://www.gov.uk/guidance/consultation-and-pre-decision-matters Accessed 16/05/2025.

Harrin, E. (2020) *Engaging Stakeholders on Projects.* Association for Project Management: London.

International Finance Corporation (2007) *Stakeholder engagement: A good practice handbook for companies doing business in emerging markets* [Online] Available at: https://www.ifc.org/en/insights-reports/2000/publications-handbook-stakeholderengagement--wci--1319577185063. Accessed 15/05/2025.

Involve (2025) *What are the quality standards for participatory and deliberative processes?* [Online] Available at: https://www.involve.org.uk/resources/knowledge-base/standards-participatory-deliberative-processes. Accessed 12/8/2025.

Kelly, R. and Kelly, U. (2019) *Community Engagement on Climate Adaptation: An Evidence Review*. Environment Agency: Bristol.

Last, S. (2019) *Communication Essentials for Business*. [OpenTextbook/RWU Pressbooks adaptation based on the original work].

Maddaloni, F. D., Meira, L. H., de Andrade, M. O., de Melo, I. R., Castro, A. and Locatelli, G. (2025) The dark legacy of megaprojects: A case of local disengagement, missed opportunities, and social value dissipation. *International Journal of Project Management* 43, pp. 1–18. https://doi.org/10.1016/j.ijproman.2025.102676

RIBA (2020) Plan of Work 2020 Overview. RIBA: London.

RIBA (2024) Engagement Overlay to the RIBA Plan of Work. [Online] Available at: https://www.architecture.com/knowledge-and-resources/knowledge-landing-page/riba-plan-of-work-engagement-overlay?srsltid=AfmBOorVD4ggacDilD1XouCnopsVrkSfhBvfD6R35mf7fBsK6xmk-uqi Accessed 05/02/2026.

Suttle, C. (2025) 'Book interview'. Interview by Simon Wilson [Microsoft teams] Unpublished, 29/7/2025.

Vangen, S. (2017) Developing practice-oriented theory on collaboration: A paradox lens. *Public Administration Review* 77:2, pp. 263–272. https://doi.org/10.1111/puar.12683

Section 2

Get ready for a change!

5 A new focus for the built environment

> There is no power for change greater than a community discovering what it cares about.
>
> (Wheatley, 2002, p.55)

Our experience focuses on built environment projects which are not housing, and where the primary purpose is functionality not necessarily beauty: flood defences, pollution treatment sites, breakwaters or weirs, to name but a few. Often these projects aim to address risks such as flooding or pollution. In our experience, and in line with what we see within the built environment generally, social value and stakeholder engagement, if considered at all, are often seen as either afterthoughts or a necessary evil. Certainly, they are viewed as less important than the technical and financial aspects of the project.

This creates both problems and missed opportunities for the project, the project team, and the field of social value.

5.1 Problems with neglecting social value and engagement

Neglecting the social element of sustainability can generate local opposition to projects, causing upset and delays, negatively affecting the reputation of the organisations involved (Fitton and Moncaster, 2022).

In August 2024, the Royal Institute of British Architects (RIBA) asked its members about planning delays:

- 19% reported projects being abandoned in the last three months.
- 35% reported projects being delayed by up to a month.
- 57% reported projects being delayed by between one and six months.
- 42% report some projects being delayed by six months or more (RIBA, 2024).

There are multiple causes of such delays, but one is local opposition and a lack of support from stakeholders and communities. It's typical of the difficulty of finding out the impacts of social issues that the specific figures are not available.

DOI: 10.1201/9781003659877-8

More subtly, who wants to be associated with a project that is unloved by the community over the long term? This can damage reputations. Above all, negative community response to one project creates a difficult environment for future projects, in particular where community objections have been ignored.

Often, we are brought in as consultants by an organisation facing the legacy of poor stakeholder relationships based on past experiences. The first thing we have to do is repair damaged relationships and that can take some time, before we even get to moving the project forward. It's better to avoid this type of problem occurring in the first place by looking at the social aspect of the project early. You wouldn't leave the assessment of the economic feasibility until halfway through the design stage, would you?

5.2 Missed opportunities from neglecting social value and engagement

The missed opportunities are more subtle, but important to the success of the project. To turn our previous question on its head, who would not want to be associated with a project that is loved and valued by the community over the long term? Where stakeholders have a positive experience of being involved in the development of a project, they are more likely to respond positively the next time.

Even if the result is not what stakeholders wanted, they are more likely to accept a project if they have had a fair chance to have a say, and clear information about why it was not possible to achieve everything they wanted. This provides a realistic base for continuing dialogue.

Failing to embed social value in project planning and implementation misses the potential to create better results. People in communities often know the place where they live well. Local knowledge can bring ideas to improve built environment projects. Residents can foresee problems that may not be captured by technical risk registers, and propose solutions from a different perspective (Anthony, 2024). Neglecting their perspective is to neglect vital knowledge that could contribute to making better projects.

Thinking back to the 'quality' component of projects discussed in Chapter 2, social value and stakeholder perspectives provide both a measure of quality, and also an opportunity to enhance the quality of projects for the benefit of society.

5.3 Why is social value undervalued?

We set out some of the background, theory, and definitions for social value in Chapter 3.

In UK legislation, social value is defined as the 'economic, social and environmental wellbeing of the relevant area' (Public Services (Social Value) Act, 2012) but the term is not well understood by those involved in designing and delivering projects in the built environment.

'Although it is increasingly broadly applied, the term 'social value' comes from the parlance of procurement' (Samuel and Watson, 2023, p.944) and the broad definition given above is difficult to square with the tools used for assessing and auditing costs and benefits of projects. In practice, how do you assess 'economic, social and environmental wellbeing' and how do you feed that into a cost-benefit analysis?

Social value is positioned as one of a number of buzz phrases – corporate social responsibility, social capital, ESG – that are also often poorly defined and fuzzy as we saw in Chapter 3. Other concepts such as environmental benefit, wellbeing, and community development muddy the waters further. Understandably some project team members are often sceptical particularly when they come across the term 'social value' in the context of procurement or compliance. Often teams pay a specialist consultant to complete their social value statement or tender question. Great news for social value consultants but confirming this as an archetypal 'tick box' issue (Fitton and Moncaster, 2019).

Our perception is that project managers for built environment projects focus on what is included in procurement requirements and project/programme management guidelines. Our main focus being infrastructure, these are often determined by public bodies.

We believe that successful built environment projects should give equal weight to these three main elements (see Figure 5.1):

- Technical feasibility.
- Social success.
- Financial viability.

This is the aspiration. What's the reality?

The technical and economic elements tend to be defined in terms of objectives and outcomes. The social, where it is present at all, is often defined in terms of processes and procedures such as defining metrics, identifying tools, and constructing an action plan. Where the social dimension is a requirement, it usually features steps in a process rather than setting objectives in terms of social acceptability, community concerns, and priorities. Again, this can lead to a 'tick box' approach where projects need to demonstrate what they have done rather than what they have achieved.

Figure 5.1 The three aspects of a successful project.

Source: Authors' own work.

Of course, the boxes need to be ticked, but for social value there is no associated legislation and little codified good practice so it can be difficult to know what should be included.

We are advocating for a shift in mindset and practice towards a focus on social objectives and outcomes, rather than going through the process of agreeing on a social value policy or a community consultation.

One way of looking at this is to see social value as the outcomes to be achieved, and stakeholder engagement as the means of achieving those outcomes. Yet stakeholder engagement is also often undervalued.

The case study below describes how a lack of focus on what mattered to the local community caused a project to stall. Later, focusing on engaging the community and finding out what mattered to them enabled a successful approach.

Case study: Lower Saâne 2050 territorial project

The River Saâne in France flows through Normandy to the English Channel. From 2000 to 2024, a major project was developed to address flood risk and related water issues.

The Lower Saâne valley had been modified in the 19th century when dykes and embankments were built to hold back the sea which was vital for agriculture and the burgeoning tourist industry. A century later, these structures were no longer effective in the face of more frequent storms and sea-level rise, leading to frequent flooding and ecological damage. The municipal campsite at Quiberville-sur-Mer, near the sea and the mouth of the River Saâne, was increasingly vulnerable to flooding. During a flood in 1999, over 100 caravans were destroyed.

The Lower Saâne 2050 project aimed to restore wetlands and intertidal habitats to support biodiversity. A new campsite was built 700 m inland on a hillside plot. The project also included building a water treatment and sewage works in nearby Longueil. Finally, the project created a new river channel reconnecting the Saâne to the sea, including building a new bridge across the river (PACCo, 2025).

A first iteration of the project was developed between 2000 and 2010. It focused on flood risk, without encompassing the economic and social dimensions. This iteration of the project ran into local opposition and was rejected by residents of three local villages in 2010. Local residents were concerned that removing the historic flood defence embankments would increase the risk of flooding to their homes.

From 2012, management of the project was taken over by Conservatoire du littoral, a French public agency whose remit is to protect and manage coastlines and the banks of lakes. Conservatoire du littoral developed a more comprehensive project approach which included environmental, economic, and social objectives. This was backed by a multi-stakeholder steering group which oversaw the project and developed the funding model.

The original hydrological studies were still relevant, but the new approach encompassed an economic dimension and, critically, sustained engagement with local communities. It took around five years to rebuild trust with local communities using this new approach. Many of the local residents were older, so a key communication tool was a physical newsletter which was seen as more likely to reach people than QR codes. Information boards were also provided, featuring different stakeholders (e.g. fisher, hunter, walker) explaining what they valued about the local area. There were two kinds of boards: information boards with an extended version of the newsletter's content and summer exhibition boards on a different topic each year: the first year was about the PACCo and Saâne projects, the second year was about fauna, flora and habitats study results, the third year was about why people love the Saâne valley, and the fourth year was a comparison between pictures from the 1950s and from 2024 (to show how much the territory had evolved). Camille Simon from the Conservatoire du littoral told us:

'The point was to get them to speak about this territory and let them tell us why "I love the Saâne valley"' (Simon, 2025).

These were backed up by online communications and public meetings. By the early 2020s, trust had been rebuilt with communities and the new approach which included economic and social objectives was widely supported.

The community had a say in some areas of the final design of the project. In particular, a landscaped path and bike trail were incorporated which provided views over the area to benefit local residents. This feature had no benefits in terms of hydrology or biodiversity but enabled people to access the area without damaging the local ecology. It required the project team to make an additional case for funding as it did not fall within the parameters of the environmental objectives (Simon, 2025).

Lessons from stakeholder engagement perspective

- The success of the project depended on systematic engagement with the community over a significant period of time.
- Stakeholder engagement methods were adapted to the particular communities.
- A tight, integrated project team was key to success.
- Stakeholders had a say in the final design of the project.

Lessons from social value perspective

- Understanding both people's concerns and aspirations was key to the project, which explicitly asked stakeholders what they 'love about the Saâne valley'.

- The original approach was unsuccessful because it did not address people's fears that removing embankments would endanger their property. As well as the fact that the main road on the coastline was removed, which could endanger the economic life of the territory.
- Access to the landscape was important to local residents.

Takeaway

Understanding both people's concerns and aspirations backed up by systematic communication enabled project success.

So where does the environment fit in all of this? Why is there not a fourth segment to our diagram? We have an open mind about how to represent the interactions, but our current thinking is that environmental considerations feature in two quite different ways:

- As a technical consideration – for instance, a requirement for a project to demonstrate biodiversity net gain or contribute to a net zero target.
- As a social consideration that emerges from engagement around a project – for instance, the value a community places on the red squirrels in local woodlands.

We believe this distinction is important, particularly because stakeholders may not value all aspects of environmental impact equally. In some cases, they may prioritise one environmental benefit over another, such as favouring water quality improvements over biodiversity, or flood protection over the conservation of rare grasslands. This highlights the need to avoid making assumptions about what people care about.

5.4 Why is stakeholder engagement undervalued?

Project team members often have experience, or hear stories, of how necessary developments are held back by community protests. The acronym NIMBY (Not In My Back Yard) has become a shorthand for objections based on self-interest such as defence of property prices or the view from the back window, against the technical benefits of a project in terms of reducing flood risk, pollution, or other environmental hazards. Political debate has weaponised and distorted these concerns, for instance pointing to objections to wind farms' effects on the landscape to support a case against renewable energy.

Project team members may have had negative experiences of stakeholder engagement, having to spend evenings having uncomfortable discussions in village halls. Stakeholder views, and in particular public views, are often experienced as not only self-interested but also unrealistic or ill-informed. The public is described as 'NIMBYs' or 'objectors' (Freewheeling, 2024). We have heard project team members talk about the need to combat 'misinformation' or 'myths'.

These concerns subsist in a context of tight project management where cost structures and timelines are sharply managed. Stakeholder engagement can be seen as time-consuming, adding little or nothing to the success of the project (Correia et al., 1998). It also often incurs substantial costs (Pagliara, El-Ansari, and Henke, 2023).

Engagement can also be mislaid among the different responsibilities of project team members. Emily Clarke from the consultancy Binnies in the UK, who has worked on numerous projects in the water sector, told us:

> I think you really need to have social buy-in for any project to work. Sometimes it gets forgotten. I can think of a project where we are providing a pipeline. The team has done all of this master planning. But there's an art to ensuring all the relevant permits and permissions are in place and despite best intentions, the timing in which this all comes together can often be disruptive to locals. For example, the project timeline may end up meaning it's not the best time of year to go through a farmer's field but delays to work around crop schedules will have a huge impact on the overall programme. It seems that overarching communication of not just why things are happening, but why they need to happen in certain timeframes is often forgotten. And this can and has really aggravated the farming community in particular for this project. Ultimately it will be called a success, because society needs it. But in the midst of all of that, you've got all of this social unrest that doesn't get captured or is left to other teams to resolve from different parts of the same organisation.
>
> (Clarke, 2025)

Statutory requirements for local authorities to conduct public consultation on planning applications before a decision is made focus heavily on the process to be followed but say little about the value or importance of consultees' views (GOV.UK, 2022), reinforcing the sense of hurdles to be overcome rather than insights to be gained.

Until quite recently much of the professional development and ongoing career development for project managers and technical project members gave little priority to issues of communication and engagement (although this is changing – see Chapter 4). One solution is to bring in specialist skills with one or more stakeholder engagement professionals joining a project. This can work very well. However, as we have already mentioned in previous chapters, in our experience stakeholder engagement professionals often feel ignored or sidelined and are often brought into the project team far too late to input into processes effectively. They may be asked to do all of the stakeholder engagement work themselves, while lacking the technical knowledge that should inform the engagement. From the perspective of other team members, this additional 'support' usually leads to more work, and this can be unwelcome to overstretched project teams. Without wishing to criticise individuals, for some organisations the skill set and objectives for their engagement specialist are framed narrowly, more focused on the 'broadcasting' aspects of getting the message out, while lacking the more extensive range of skills required to create and sustain a context in which stakeholder views are sought and used to improve and enhance projects.

5.5 Towards a new focus

This book is all about doing things differently and disrupting the norm, so what would good look like?

We believe that there is a major opportunity to work with stakeholders to understand social value and create better, more socially successful projects in the built environment.

If project managers and project team members give social value equal billing to the technical and financial aspects, they can deliver projects that do their job well and are appreciated, even loved, by the people who work with them or live near them, as well as hopefully generations to come. The emotional legacy can be a strong positive.

This can be achieved by integrating the skills and methodology of stakeholder engagement at every stage of a project.

The skills for great engagement need parity of esteem with financial and technical skills and knowledge – indeed, these are a knowledge domain and professional skillset in their own right.

The methodology required is described in different ways by different sources but includes a focus on understanding who stakeholders are and how to reach them, including clarity about what they can and cannot influence. Later chapters in this book step you through this process.

Above all, it's about shifting mindset so that rather than being a burden or obligation, the social dimension is understood as a key to doing better work and building better projects – and often saving time and hassle along the way.

This new focus will bring some disruption as it requires project managers and project teams to state engagement objectives and listen to stakeholders early in projects, valuing the social dimension on equal terms with the technical and financial dimensions.

As this chapter outlines, we believe that disruption is worth it because it will lead to better results. The next chapters provide guidance on how to make the change with minimum disruption and effort, but maximum results.

Key takeaways

Great projects should aim to be technically feasible, financially viable, and socially successful.

Social value and stakeholder engagement can help achieve this, but they are both often undervalued.

We need to shift the mindset of project teams, so the social dimension is seen as central to great projects. Otherwise, projects may face opposition, delays, budget constraints, and reputational damage.

References

Anthony, B. (2024) The role of community engagement in urban innovation towards the co-creation of smart sustainable cities. *Journal of the Knowledge Economy* 15, pp. 1592–1624. https://doi.org/10.1007/s13132-023-01176-1

Clarke, E. (2025) 'Book interview'. Interview by Simon Wilson [Microsoft Teams] Unpublished, 17/6/2025.

Correia, F. N., Fordham, M., Saraiva, M. D. G. and Bernardo, F. (1998) Flood hazard assessment and management: Interface with the public. *Water Resources Management* 12, pp. 209–227. https://doi.org/10.1023/A:1008092302962

Fitton, S. L. and Moncaster, A. M. (2019) Arguments for a co-production approach to community flood protection. *Proceedings of the Institution of Civil Engineers – Engineering Sustainability* 172:7, pp. 335–344. https://doi.org/10.1680/jensu.17.00014

Fitton, S. L. and Moncaster, A. (2022) Social value, infrastructure and stakeholder engagement: A complex triangle. *Proceedings of the Institution of Civil Engineers - Engineering Sustainability* 175:4, pp. 194–201. https://doi.org/10.1680/jensu.21.00030

Freewheeling (2024) *On time, on budget - and ast* [Online] Available at: https://www.freewheeling.info/blog/-on-time-on-budget. Accessed 14/05/2025.

GOV. UK (2022) *Guidance: Consultation and pre-decision matters* [Online] Available at: https://www.gov.uk/guidance/consultation-and-pre-decision-matters. Accessed 14/05/2025.

PACCo/Interreg *Basse Saâne 2050: Dossier de presse* [Online] Available at: https://basse-saane-2050.com/. Accessed 12/07/2025.

Pagliara, F., El-Ansari, W. and Henke, I. (2023) 'A methodology to estimate the benefits and costs of stakeholder engagement in a transport decision-making process' in *Smart and sustainable built environment* 14:2, pp. 326–341. October 2023. https://doi.org/10.1108/SASBE-03-2023-0049

Public Services (Social Value) Act 2012 (2025) [Online]. Available at: https://www.legislation.gov.uk/ukpga/2012/3. Accessed 04/05/2025.

RIBA (2024) Future Trends Survey: August 2024.

Samuel, F. and Watson, K. (2023) Social value of the built environment. *Buildings and Cities* 4:1, pp. 938–947. https://doi.org/10.5334/bc.399

Simon, C. (2025) 'Book Interview'. Interview by Simon Wilson [Microsoft teams] Unpublished, 27/5/2025.

Wheatley, M. (2002) *Turning to One Another: Simple Conversations to Restore Hope to the Future.* Berrett-Koehler: San Francisco.

6 A new approach to project design

> Design shapes the world. It has huge power and with that comes responsibility. We can design things in, and we can design things out. The more intentional we are about this, the more design can make life better for all.
>
> (Design Council, 2022, p.1)

Building something changes a place and changes the lives of the people who live, work, and play there.

'Iconic' buildings like the Guggenheim Museum in Bilbao have the power to transform neighbourhoods but the same goes for small, satisfying projects doing their job well like a community café, a bridge, weir, or breakwater. The same goes for buildings and structures that people dislike. People feel this viscerally when a building or structure goes up that they hate. It changes how a person feels about a place by the way it looks, maybe sounds or smells, and also the way it affects how people behave, their sense of wellbeing, and place. In his provocative book *Humanise* (2023), Thomas Heatherwick explains how the poor or 'boring' design (as he refers to it) of a building and even cities can be detrimental not only on a personal level but on a societal level (Heatherwick, 2023, p.141):

> 'Boring places make us stressed.
> Boring places make us sick.
> Boring places make us lonely.
> Boring places make us scared.
> Boring places contribute to division and conflict.
> Boring places are unsustainable.
> Boring places are unpopular.
> Boring places are unfair.'

This chapter seeks to present, and then unpack, a new approach to project design: one that puts the stakeholder at the centre and incorporates the social and emotional aspects that enable projects to be socially successful. This approach, with social value at its core, underpins the methods to strengthen your projects set out in Chapter 7 onwards.

DOI: 10.1201/9781003659877-9

6.1 The social and emotional impacts of built environment projects

At the beginning of the century, the Commission for the Built Environment (CABE) was one of the first organisations in the UK to categorise the different types of value we find in the built environment (CABE, 2006). They expanded the concept from just financial, environmental, and technical value to:

- **exchange value:** the value the commodity has to be traded.
- **use value:** the value gained from the building or structure being used.
- **image value:** the value of an innovative design or the image of the design.
- **social value:** the positive values created for those stakeholders impacted by the design and project.
- **cultural value:** the contribution, the design, or development will make to the cultural heritage of a place.

Looking at social value, CABE focused on 'how places and buildings encourage people to interact in ways which lead to trust, mutual understanding, shared values and supportive behaviour' and how social value encourages community spirit, people feeling proud of their neighbourhood and safe to go out at night, and better mental and physical health (CABE, 2006, p.25). Relating the social value perspective to the built environment was new thinking at the time, and CABE was one of the first organisations to see the importance of social value, six years before the Public Services (Social Value) Act 2012 was passed.

As we saw in Chapter 3, industry focus on social value today is broader and includes some of the factors CABE (2006) categorised as image, environmental, and cultural value. In 2011, CABE was incorporated into the UK Design Council which produced a Design Value Framework (2022). A similar approach is found in the UK Green Building Council's Social Value Framework (2022). These factors affect how people feel about where they live, work, and play. The Royal Institute of British Architects (RIBA) suggests that these feelings are often what matters most to people. RIBA's Value Toolkit examines the social value of design through positive emotions, connections, participation, freedom, and flexibility (RIBA, 2020).

However, these factors are intangible, difficult to measure, and frequently missed in project appraisals as they are difficult to include in quantitative project justification processes such as Cost-Benefit Analysis. These values go deeper than just the aesthetics or 'image value' of an individual structure. They encompass deeper feelings about change and authority in a person's life.

6.2 How social values can deepen resistance to change

We often hear (and probably say) 'people don't like change'. Building something new is a fundamental change to a person's surroundings or daily routine, and people often don't like it. Resistance is often collective as people in a community come together – a positive act of collaboration which is experienced by professionals as negative. Resistance to change can seem irrational. It is sometimes hard for professionals to understand why stakeholders don't want, say, a breakwater when they have the knowledge and data to justify why it is needed and can envisage the benefits it will bring in the long term.

There is evidence that a major factor in resistance to change in organisations is deep-rooted values and emotional loyalty to how things are currently: the status quo (Pardo de Val and Fuentes, 2003). Change can lead to the unknown and most people are not willing to risk the unknown in case their experience is worse than the current situation. In the case of the built environment, we need to understand what people value about a place and how those values might be affected by building something. However well thought through and well-funded, development *can* negatively impact stakeholders and communities, no matter how much money has been invested in it. A breakwater may do an excellent job of reducing the risk of tidal surges but, however well landscaped, spoil the view that residents enjoy when walking their dogs every day or running their chosen route.

It's not that people 'don't like change'. They tend to resist change if it offends their values, if they do not see a reason for the change, and if they feel that change is being done *to* them rather than *with* them.

6.2.1 People tend to resist change if it does not align with their values

People's perceptions of the place they live, work, or play are filtered through personal experience, history, and culture. This is amplified when people in a community come together to protect what is important to them. They often use historical and cultural factors as justification for resistance, for example:

- 'This is a quiet market town, and we don't want it disrupted'.
- 'This place has quite a buzz and there's plenty going on and we don't want to lose that'.
- 'People have been farming here for generations'.

Statements like these express the deep value that people place on history, continuity, and their reluctance and sometimes fear of disruption, even when there is a strong technical case for change.

6.2.2 People resist change if they don't understand why it is necessary

Often opposition comes from stakeholders not buying into the reason for a project, for example

- 'Why do we need a new bridge anyway?' – when you haven't explained that the old bridge is now unsound and a safety risk.
- 'No one's ever worried about the pollution before' – when you haven't explained the new evidence of harm and the health implications to the community.
- 'We've not been flooded here, so we don't need any unsightly structures' – when you haven't explained the impacts of climate change and future flood risk predictions.

Engagement often begins with an assumption that the reason for change is well understood and shared. If that is not the case, there is little point in asking people how they want to see a project developed, if they are not persuaded why it is needed. Dr Jacqui Cotton, Flood Risk and Coastal Change Research Manager in the UK's Environment Agency, explained how 'the success of a project rests on clearly setting out and communicating the project objectives and outcomes'. If the aims, objectives, and outcomes are not identified by the project team and then clearly communicated to the stakeholders involved, perspectives will be different, and expectations will not align. Consequently, different people will have different perceptions of what the project is for, which in turn will lead to difference of opinion and subsequently conflict and opposition.

In the case study below the community felt that they were not being listened to and their values ignored.

Case study: Didsbury flood storage improvements scheme

The Didsbury flood storage basin is 62 ha and used to relieve peak flow on the River Mersey. Didsbury is in Greater Manchester, UK, and the flood basin is predominantly used for recreational activities with footpaths and cycle paths, a rugby club, allotments, and a small number of residential properties within the vicinity. When the basin is in use and storing water, some of these facilities can be impacted by floodwater. An improvement scheme was commissioned by the Environment Agency in 2009 due to it not performing as required. The proposed improvements consisted of:

- Installation of new monitoring and telemetry equipment.
- Construction of a floodwall and a floodgate to protect the residential properties.
- Installation of flood mitigation measures in the local amenities.
- Creation of a maintenance strategy to ensure no obstructions impede the flow of floodwater (Fitton and Moncaster, 2019).

This case study was used as part of Sarah's PhD research (Fitton, 2015).

In 2009, a contractor was appointed by the Environment Agency to design the scheme. They designed the scheme to the required specification and once the designs were near completion, they went out for consultation with the local community. Unfortunately, the designs were not to the liking of the local community. The local community were very opposed to the scheme and organised themselves effectively to create a lot of opposition, including contacting the media and local MPs. The opposition was so fierce that the Environment Agency had no choice but to halt the project, change contractors, and start again on a new design.

During Sarah's research, she spoke to local community members to try and understand why the opposition was so strong to the initial design. She found two main reasons: the local community felt they weren't listened to or consulted due to the design being near completion when they first found out about the scheme. The second reason was that the scheme was not aligned with the social values of the community.

Local community members voiced how they felt they were being told what the scheme was, rather than having any influence or ability to shape how the scheme would progress. They felt their views weren't considered or even seen as important given that they heard about the scheme so late in the project process. This created distrust with the contractors, and the local community wanted no dealings with them. There was essentially no relationship.

One of the main areas of contention was due to what the contractor had proposed for the allotments. In order to meet the required project brief, they were proposing to do work on the local allotments, which would result in the loss of a significant proportion of them. Sarah spoke to a local resident and an allotment holder to find out why there were such strong feelings. The resident explained to her that their allotment was very important to them. They lived alone and were retired. They got up at 6am each morning to tend to the allotment and then went back in the afternoon. They explained that if they didn't do that, they had no reason to get up in the morning. Also, the only social interaction they had some days was with the people they saw at the allotment. Therefore, the issue of the allotment was much deeper than just a recreational space; this was something that was intrinsic to this person's social values and quality of life.

The opposition was so strong from the local community that a new contractor was appointed and they were directed to leave the allotments as they were and to create a new scheme. This was at an additional cost to what had already been designed, not to mention the delay to the programme.

The new contractor told Sarah that when they started the project, they were met with hostility from the local community which proved difficult. However, they worked hard to repair the relationship and ensured they included the local community in decision-making.

The result was a new design that met the brief, left the allotments alone and was acceptable by the Environment Agency as the client and the local community.

Lessons from stakeholder engagement perspective

- Stakeholder opposition can derail your project, fast.
- Trust is a very fragile thing – break it and it takes a while and a lot of hard work to rebuild it.
- It is easier in the long run to engage with stakeholders and deal with their concerns as the project progresses than try to deal with their concerns (and their frustration) all at once.

Lessons from social value perspective

- Projects can not only create social value, but they can also destroy it.
- Local knowledge and engagement are the only ways to understand stakeholder social values.
- Ignore stakeholder social values at your peril.

Takeaway

Early stakeholder engagement on a project is important to develop relationships and build trust. It is also important in facilitating the identification of social value: not just social value that the project can create, but the social value of the existing site (Figure 6.1).

Figure 6.1 The River Mersey flowing through Didsbury.

Source: Fitton (2015).

It's important to recognise that the community may be right. Sometimes the justification for a project is weak, and stakeholders are pointing out what the project team can no longer see as they are too far down the track of design. Just because there is funding for a project team, a client willing to deliver it, and the political will for it to proceed, does not mean it is what the community or stakeholders want or need.

6.2.3 People resist change if they are not involved in the decision-making process and not able to have their say

We all recognise how it feels when things are done *to* us rather than *with* us. In the UK, groups advocating for disability rights use the phrase 'nothing about us without us', referring to a legacy of imposed decisions which left people feeling disempowered.

How decisions are made is as important as what decisions are made. Even if there is strong evidence of which stakeholders will benefit from a change, failing to involve them in the decision-making risks opposition.

6.2.4 People resist change because they don't trust the decision-makers

In the years we have worked with projects that affect the public, distrust has been a rising tide. It may be most obvious in relation to politics and politicians, but distrust affects most authority figures and communities will see industry professionals as authority figures. Communities often feel, and are, disempowered when authority figures tell them their plans, even those supposedly benefiting the community, but they are not given a voice or listened to (McEwan et al., 2022).

People perceive politicians and others as untrustworthy because they believe those people display attributes such as unfairness, acting in their own interests rather than the interests of those they are supposed to represent, and incompetence (Bertsou, 2019).

For the UK, Brexit offers an example of this. Following the results of a referendum held in 2016, the UK officially left the European Union (EU) in January 2020. The case for Brexit was largely based on distrust of the EU institutions which were seen to treat the UK unfairly. People who voted against Brexit felt that promises made during the referendum were false and believed that certain politicians wanted to persuade voters to vote to leave the EU due to their own self-interest. Since the UK left, another narrative claims that the failure to secure 'Brexit benefits' was down to the incompetence and self-interest of politicians. The same set of events deepened distrust on both sides of the debate.

This is just one example of how trust in the political system and authority has been eroded and why stakeholders, communities, and the public are sceptical when new developments are proposed.

Stakeholders may distrust your plans because they see them as unfair:

- 'Your plan will only benefit people who already own a home, not us tenants.'

  They may believe that you are motivated by self-interest or vested interests:
- 'You're only doing this because big business wants it.'

  Or, very importantly, they may doubt your competence:
- 'Have you seen the mess they've made of the estate down the road? Why would you do a better job than they have?'

Trust from stakeholders has to be earned not assumed. Resistance to change and distrust are easy to dismiss as 'irrational', and it is tempting to try to shut down the emotional response and calm it with a barrage of facts. However, as we have seen, this response by communities and stakeholders is not necessarily unjustified. And it's more effective to work with the grain of social value than against it.

6.3 Working with the grain of social value

Understanding the values your stakeholders hold, and the emotional impact new development can create, should be considered in the early stages of a project. For the project to achieve as much socially as it does technically, stakeholder values and emotions need to be seen and taken seriously.

It is important to have and show empathy for stakeholders' views and perceptions. You need to understand why they may be wary, why trust may be low, and why they may fear or be sceptical about change. Work with the grain of values and emotions to understand where they come from. This can start to create a more positive relationship. Demonstrating competence, fairness, and a broader interest (rather than self-interest) can reinforce that positive start.

There are a number of ways for project teams to work with the grain of stakeholders' social value to benefit their project. The different skills and attributes are shown in Figure 6.2 and explained in more detail in this section.

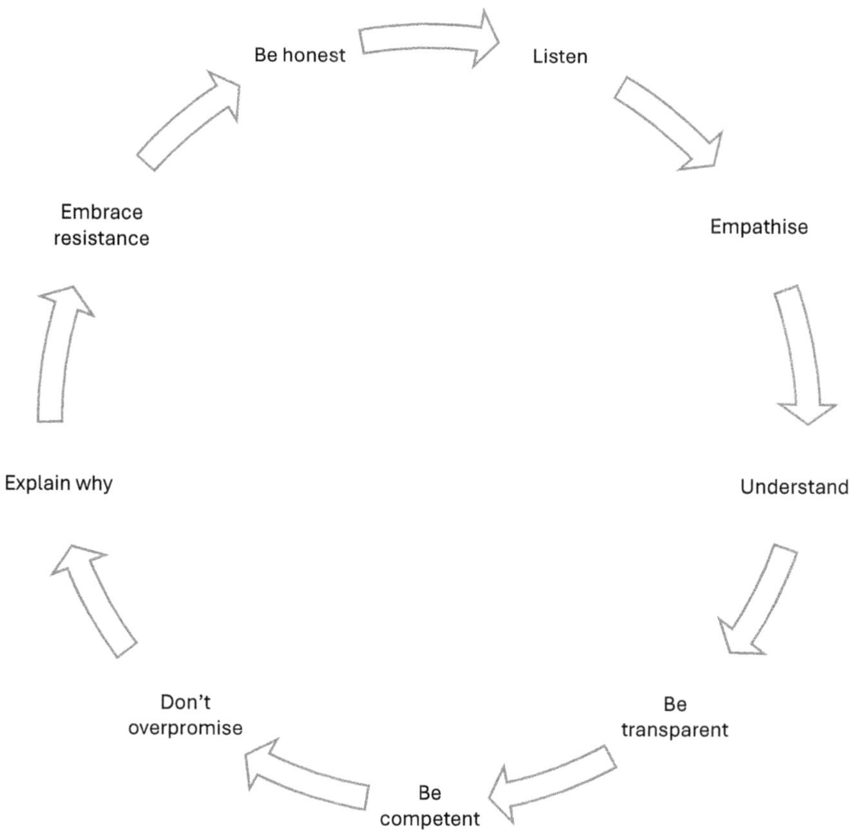

Figure 6.2 Skills and attributes required to work with social values.

Source: Authors' own work.

1 *Actively seeking to understand the social values associated with the place and your project*

Start by seeking out people and groups who can help you understand the place and what they value about it. In Chapter 9 we write more about who that might be. The key skill here is listening without judgement and with compassion to those who want to speak and those you want to speak to. Hearing the way individual and collective values are expressed will deepen your understanding of what is important.

Timing matters. There is no point listening to what people want when the conceptual design is 90% completed and the project budget is dwindling. Giving stakeholders the chance to put forward their thoughts should be scheduled early in the project process alongside context setting and early feasibility studies, with a realistic budget and timeline.

Findings from these conversations and the thoughts and opinions heard can help authentically shape the design going forward. This doesn't just mean the colour of a wall, or the type of plants used to enhance the aesthetics, but the foundational concept of the design: how it will function, how it will look, how it will impact people on a daily basis.

How you respond to what stakeholders tell you sets the tone for building trust. It is tempting to respond, even justify, if you hear difficult things about your project. As professionals we know how a project should technically work, what legislation it has to adhere to, what standards it needs to meet, and the economic case needed for it to be approved. Most stakeholders – be it a community, the wider general public or, in the case of a large infrastructure project, society as a whole – do not have this knowledge. At this stage, your role is listening to understand, not listening to respond. Managing expectations is important, but the skill of listening is key both to gaining insight into social values and also demonstrating that you can be trusted to listen without judgement.

2 *Being open to, and acknowledging, emotion*

During these conversations, people often express strong emotions, positive or negative, about a place or project. You may find hearing these emotions and the way they are delivered difficult – we often do. In UK culture at least, professionals are used to cloaking their real thoughts in diplomatic language and an equable tone: the need to stay professional at all times. It can seem difficult to find relevance in what you are told at this stage, when emotions are heightened, and you are confronted with statements like:

- 'There's just something about the place.'
- 'It's just got something special, and it would be a shame to change it.'
- 'There's nothing here for us young people – I'm going to have to move away after school or university.'
- 'It's a place for the kids to play and there's nowhere else local.'

Empathy is a key skill. Flood risk management is a good example of this. During flood risk management projects, the stakeholders who are most impacted will usually be those who have suffered the effects of flooding, whether a single occurrence or on multiple occasions. The emotional impact of flooding and the mental health implications of living with the risk of flooding are significant (Public Health England, 2020). Consequently, conversations with these stakeholders may be highly charged and emotional. Professionals need to have the empathy to understand the context and circumstances of the stakeholders and the often-emotive responses, in order to gain the most insights.

This is by no means easy. In emotional situations, people may look for someone to blame or want to ensure that no one else has the experience they have had, and often professionals bear the brunt. You, or people like you, may be seen as having caused the issue or having failed to prevent it. Having the empathy and the time to listen to stakeholders, then reflecting people's comments back can help clarify what really matters to them. This helps develop a long-term relationship in which trust is crucial.

3 *Demonstrating trustworthiness*

> And where there is trust, people work together more easily and effectively.
>
> (APM, 2025)

All human relationships are built on trust. If there is no trust, 99% of the time, the relationship will not last. This goes for the relationships with your stakeholders. There needs to be trust for engagement to work effectively (Leyden et al., 2017).

Above we mentioned research suggesting that people are distrusted if they show incompetence, unfairness, or are seen to act in their own interests (Bertsou, 2019). So, it helps to build trust if you act in a fair- and open-minded way and show that you are competent.

Start the habit of openness early in a project. Admit what you know and what you don't know. Be human. Start the process of listening early and take the stakeholders on a journey with you, don't go on the journey for them and tell them what they got at the end. Trust needs to be earned, and this takes time. You can't expect to start speaking with stakeholders and immediately build a trusting relationship. Just like personal relationships, the first few meetings are quite formal and neither of you give too much away, but as you get to know one another, you become more trusting of each other. Starting to develop relationships with your stakeholders early creates potential for a more solid and trust-based relationship to develop over the longer term.

Competence also builds trust. If you say you will do something, make sure you do it. Follow through on tasks and on deadlines ('You said, we did'). Demonstrate your competence in everything you do. Communities and stakeholders may be sceptical of professionals due to the lasting legacy of other built environment projects. Sometimes we are already on the back foot before we have even started the project! It is easy to erode trust by not doing what we promised. This can range from the smallest thing such as providing project information or speaking at a local meeting, to more important aspects like promising to save an important community building or supporting a local charity.

Don't over-promise, manage expectations, and be realistic as to what you and the project team can achieve. This is something Jo Dobson, Director at Useful Projects, champions. She explained that on projects it's important not to promise the earth if the project has no intention of delivering it, just to keep people on side. It will backfire in the long run. Doing what you say will contribute to the development of that trust that is so critical in stakeholder relationships (Dobson, 2025).

Finally, build trust by trusting. Show that you trust stakeholders to know what they are talking about. Don't pour doubt on their concerns but work with them to build a common understanding of what matters for the project.

4 *Involve stakeholders in understanding the* 'why' *of your project*

You need to speak to stakeholders … but *after* you have listened, the sequence here is important. Listen to understand their perspective on the situation so you can formulate a problem statement that meets both the technical and the social needs. It is important to start with 'why', involving stakeholders in the problem to be solved rather than jumping to engagement on the solution that you perceive is needed to a problem that you perceive is there.

It's even possible that these 'why' conversations can lead to new insights or new solutions given the importance and richness of local knowledge (see Chapter 9).

There is obviously a need for your project, or you wouldn't be embarking on it. However, that isn't a socially acceptable reason to go ahead and impose on a community a development that they don't understand and don't think they want. If stakeholders understand why something is needed, even if they don't like it, they will be more inclined to work with you and support the outcomes rather than opposing something that doesn't make sense to them. As we discussed in Chapter 3, you will have a Social Licence to Operate.

It helps if you can work with stakeholders to link the 'why' of a project to their social value:

- 'It's important to us that our community is well connected so we need a bridge that will keep us connected for the long term.'
- 'We want our children to grow up healthy, so we need to act now to protect them from long term health risks like pollution.'
- 'It's important to us that our town is a safe place to live so we need to plan for future risks like flooding even if they aren't a threat today.'

5 *Not fearing resistance*

Resistance will arise. It's only a sign of failure if it takes you by surprise. Be prepared for resistance. You may as well try to work with it rather than against it.

Resistance is often linked to diversity and difference. In organisations, teams that see difference as a resource can challenge the usual ways of doing things and a confident organisation can see resistance as an opportunity to do things differently (Gagnon and Collinson, 2017).

Similarly, for your project, stakeholder resistance is not necessarily something to fear or 'manage away'. It can point to different ways of doing things, which in turn can benefit both the stakeholders and the project team. It helps to listen to what is being said, as well as the tone of voice. Vehemently expressed negative views may still prove useful once you understand what lies beneath the surface.

6 *Showing stakeholders that they can influence what you decide*

This is where transparency becomes most important. Stakeholders appreciate knowing that they will have an influence on decisions about a project, but it has to be real.

Stakeholders are often sceptical – 'You'll decide what you want to do whatever we say' is a common response during engagement events. The perception that the project design and development is already a 'done deal' is prevalent. As mentioned above, they may suspect self-interest or vested interests at play: 'You're doing this because you're in the pocket of big business or developers'.

In Chapter 4 we discussed the different types of engagement (inform, consult, collaborate). Employing a collaborative or co-production approach can be hugely beneficial (Stewart, 2024). Where there is scope for collaboration, you not only build greater trust and commitment, but potentially benefit from views from stakeholders who may become allies. This in turn can result in a more socially successful project with a positive lasting legacy, and also potentially reduce opposition, delay, and costs to projects.

Sometimes project teams fully intend to collaborate and co-produce a design, but it never really happens. Collaboration is not easy, and it is not a quick process. If you tell stakeholders they will have a say, you need to follow through with this promise. Don't over-promise. If the truth is that there is little scope for stakeholders to influence the design of your project, be honest about it. If you are only engaging on certain aspects, be clear about what is in scope and what is out of scope.

6.4 How to secure the benefits of social value for your projects

The fundamental aim of using social value to shape design is to achieve better outcomes. Not every design or project can achieve all the types of values CABE (2006) identified but one or more of those benefits – whatever is important to your stakeholders – can be part of the fabric of your project design.

Working with social values through effective stakeholder engagement can secure significant benefits for your projects.

At a minimum, working as we outline above to build trusting relationships with stakeholders makes it less likely that you will face sustained opposition, challenge, and delay to your project. This also has significant financial implications for your project (Bice, Neely, and Einfeld, 2019).

The more you work with your stakeholders and the community, the more you will learn. They may have technical knowledge, but more likely they will bring different types of knowledge about history, culture, connections, or relevant local issues that will help you manage risks and enhance your project.

Positive feedback from communities can enhance your reputation and brand. We have spent years of our professional lives working on projects suffering negative effects from previous community experiences. If your current project is a success in the eyes of the community, you will start the next one on a good foundation.

In the following chapters, we look at the practical steps you can take to incorporate social values in your project design and realise the wider benefits:

- Engage early, listen, then talk.
- Situational analysis and scenarios.
- Reimagining stakeholder mapping.
- Optioneering.
- Engaging through the project lifetime.
- Monitoring, evaluation, and learning.

Key takeaways

Thinking about social value helps us understand the wide range of benefits built environment projects can bring and enables us to tune in to what matters to stakeholders. Equally, projects may experience resistance and distrust if they do not respect people's social values.

Early in your project, spend time listening to stakeholders to build relationships.

Demonstrate and earn trust, be open and honest, and manage expectations.

Working with the grain of social value can reduce opposition and delays, create better results (and enhance your reputation).

References

APM (2025) *10 Key principles of stakeholder engagement* [Online] Available at: https://www.apm.org.uk/resources/find-a-resource/stakeholder-engagement/key-principles/?gad_source=1&gad_campaignid=23101617584&gbraid=0AAAAADDDFbjajZcR_jcvPtd0Nwa-GptJE&gclid=Cj0KCQiAnJHMBhDAARIsABr7b8586hqWBoMdRqYx0dRTF95hpOlczEtPY1IHhOhnd4pQ0Evkv5Km_gcaAvE5EALw_wcB Accessed 05/02/2026.

Bertsou, E. (2019) Rethinking political distrust. *European Political Science Review* 11:2, pp. 213–230. https://doi.org/10.1017/S1755773919000080

Bice, S., Neely, K. and Einfeld, C. (2019). Next generation engagement: Setting a research agenda for community engagement in Australia's infrastructure sector. *Australian Journal of Public Administration* 78:2, pp. 290–310. https://doi.org/10.1111/1467-8500.12381

Commission for Architecture and the Built Environment (2006) The Value Handbook.

Design Council (2022) *The design value handbook* [Online] Available at: https://www.designcouncil.org.uk/our-resources/the-design-value-framework/. Accessed 29/01/2025.

Dobson, J. (2025) 'Book Interview'. Interview by Sarah Fitton and Simon Wilson, [Microsoft teams], 27/02/2025.

Fitton, S. L., (2015) Social value in practice: A case of flood alleviation schemes. *PhD Thesis*, University of Cambridge

Fitton, S.L. and Moncaster, A. (2019) Arguments for a co-production approach to community flood protection. *Proceedings of the Institution of Civil Engineers – Engineering Sustainability* 172:7, pp. 335–344. https://doi.org/10.1680/jensu.17.00014

Gagnon, S. and Collinson, D. (2017). Resistance through difference: The co-constitution of dissent and inclusion. *Organization Studies* 38:9, pp. 1253–1276. https://doi.org/10.1177/0170840616685362

Heatherwick, T (2023) *Humanise: A Maker's Guide to Building Our World.* Penguin Books: London.

Leyden, K. M., Slevin, A., Grey, T., Hynes, M., Frisbaek, F. and Silke, R. (2017) Public and stakeholder engagement and the built environment: A review. *Built Environment and Health* 4, pp. 267–277. https://doi.org/10.1007/s40572-017-0159-7

McEwan, L., Holmes, A., Cornish, F., Leichenko, R., Guida, K., Burchell, K., Sharpe, J., Everett, G. and Scott, M. (2022) Rebuffing the "hard to reach" narrative: How to engage diverse groups in participation for resilience. *Journal of Extreme Events* 9:2&3, pp. 1–14. https://doi.org/10.1142/S2345737622500038

Pardo del Val, M. and Martinez Fuentes, C. (2003) Resistance to change: A literature review and empirical study. *Management Decision* 41:2, pp. 148–155. https://doi.org/10.1108/00251740310457597

Public Health England (2020) *The English national study of flooding and health.*

RIBA (2020) *Social value toolkit for architects* [Online] Available at: https://www.architecture.com/knowledge-and-resources/resources-landing-page/social-value-toolkit-for-architecture?srsltid=AfmBOop4Kot_SxIoHfhX6eIFHohbMNImhFo_eS_qVVHigUkUqGl9mQ_Y. Accessed 29/01/2025.

Stewart, I. S. (2024) Advancing disaster risk communications. *Earth-Science Reviews* 249, pp. 1–14. https://doi.org/10.1016/j.earscirev.2024.104677

UK Green Building Council (2022) *A guide for delivering social value on built environment projects* [Online] Accessible: https://ukgbc.org/resources/a-guide-for-delivering-social-value-on-built-environment-projects/. Accessed 29/01/2025.

Section 3

The practical bit – approaches to authentic stakeholder engagement and social value

7 Engage early, listen, then talk

> The most basic and powerful way to connect to another person is to listen. Just listen. Perhaps the most important thing we ever give each other is our attention.
>
> (Remen, 2006, p.143)

Key steps

- Focus on your own and others' psychological safety.
- Engage your empathy.
- Consider equality, diversity, and inclusion (EDI).
- Remember Engage, Deliberate, Decide (EDD).
- Explore whether a facilitator could help.

Despite our best intentions sometimes we feel that engagement is like calling into a cave and getting no response.

The people we most want to hear from sit on their hands, and sometimes we feel that we are contributing to the problem by being nervous and uncertain.

Everything starts with feeling safe. Of course, this includes making sure that everyone is physically safe or their attention will be elsewhere. No one would run the risk of holding an engagement event in a physically unsafe space.

It's equally important to consider your own *psychological* safety and that of your stakeholders. In this chapter we build on the approach we outlined in the last chapter and examine how we can create a space that facilitates listening and talking which is inclusive for everyone.

7.1 Building psychological safety

Psychological safety means feeling that you can speak up and express a view without the risk of being criticised, put down, or humiliated. The idea of psychological safety first emerged in the context of workplaces and was used to describe how far

DOI: 10.1201/9781003659877-11

workmates felt safe to engage in 'interpersonal risk taking' such as sharing personal information (Edmondson, 1999). At work, supportive leadership, teams, and culture as well as relationship networks contribute to psychological safety. A psychologically safe environment enhances team performance, learning, innovation, and creativity (Newman, Donohue, and Nathan, 2017). Typically, it takes time to build psychological safety for a group.

We believe that psychological safety is also critical for effective engagement. If stakeholders do not feel that they can safely express their views they self-censor, go with the majority view even if they don't agree with it, or keep quiet. All of which you don't want for your project. It is a waste of time for the stakeholders and the project team. Engagement without psychological safety runs the risk of providing incomplete or misleading evidence.

Many factors may cause stakeholders to feel unsafe including:

- Feeling they lack expertise.
- Experience of discrimination or unfair treatment.
- Previous negative experiences of engagement.
- Fear or trauma associated with the subject matter of the engagement, for instance previous experience of flooding.

In the workplace, organisations and teams can build psychological safety over time. Often engagement is brief and fragmented, allowing little time to get to know people and build the relationships that help them feel safe.

7.1.1 Step 1: It starts with you

Take time to reflect, individually and in a team, recognising that if you work within an organisation the culture and climate will affect this. If people in the organisation do not feel psychologically safe, they will transmit this to others.

It's helpful to open up this discussion by exploring

- When do you feel safe and when do you not?
- What changes when you do not feel safe?
- Who and what can help you?

In our experience, important factors include age and confidence, previous experience of having views dismissed, or a sense of the discussion not being for 'people like me'. In a citizens' panel one participant said, *I didn't want to speak up because other people know more than me.* It takes time and energy to create an environment in which that person feels able to speak up even though others may know more than they do.

7.1.2 Step 2: Engage your empathy

Put yourself in the shoes of participants. As yourself, *how would I feel if I was at a meeting where everyone was older than me/knew more than me/was an older white man/looked down their noses at the likes of me?*

Recognise that despite your best intentions you may come across as intimidating, and activities that are familiar to you such as asking people to introduce themselves, sharing PowerPoint slides, or a Q and A session may be intimidating to people who are not used to them.

The concept of neurodiversity means that people access and process information in different ways, so what works for one person may not work for another. You may find different responses emerging in a group in response to different processes.

7.1.3 Step 3: Find ways to help people settle down and feel at home

This can be quite literal. Inviting community members to meet on 'official' premises can cause anxiety. Where possible, go to where they are: places that are familiar and unthreatening.

It also means taking time when people first arrive or at the beginning of a meeting to find their bearings, take a breath, and build the confidence to contribute. We live in a world where people feel that they don't have enough hours in the day, but it's important to create time for people to centre themselves and become acclimatised to the space and situation.

It's the simple things: refreshments and small talk before starting a formal session can help put people at ease. The widely used phrase 'connection before content' encourages you to invite people to introduce themselves and make a human connection before diving into the content of the discussion.

Dr Catherine Queen, a Senior Lecturer in Planning at the University of Liverpool, UK, told of her experience working on an 80 km road upgrade in the North of England (Queen, 2025). The engagement approach was very much focused on place specificity. The engagement team working on the project were required to live within an hour's drive and have similar accents, to help develop a connection with the communities. Landowners and farmers were a large stakeholder group, and Dr Queen told us how the team worked to connect on their level. One way was to use familiar language and Dr Queen found out that farmers name each of their fields and spent time learning the names of the fields to make conversations easier.

7.1.4 Step 4: Explain why voices matter

Take time and care to explain how you will engage people, whether in a one-to-one conversation, a drop-in session, a meeting, or using an online platform. Be explicit about why and how you value stakeholders' contributions and make clear that you are not expecting them to be experts in the technical aspects of your project (unless they are). Managing expectations from their perspectives and yours is very important here.

The concept of 'expertise by experience' or 'lived experience' can be useful to help stakeholders understand why their voices matter. They may be the people who live in a place, who will be affected by the project, and their deep understanding of what happens day to day can be vital. We worked on a project looking at what to do about a worn-out weir on a river in England. Scientific analysis and modelling were complemented by the insights of one local resident who had been taking his boat out on the river for 50 years. Stories like this can help people feel at ease.

Where stakeholders take the initiative, for instance in supporting or opposing a change in their local environment, they are often assisted by technical experts to make their case. Similarly, the technical leads in a project can benefit from the particular expertise of stakeholders, working together in a participatory inquiry (Fischer, 2000).

7.1.5 Step 5: Listen and value what you hear

As professionals we may feel that professional knowledge is the only knowledge that counts (Scott, 2001). Lived experience brings different types of knowledge around local history and culture, social impacts of previous projects, risk, and opportunity as perceived in local communities (Babaei, 2023).

It's important to listen, without judgement initially, and challenge yourself to take seriously what may seem apparently trivial. We were involved in a project for the UK Department for Education which – an innovation for the time – brought children and young people into the department to talk about their education. The first thing they wanted to talk about was toilets. The officials leading the project were initially dismissive. As they listened, they heard children's concerns that toilets were unsafe spaces where they felt at risk. This line of discussion opened an unexpected perspective on safety, in this case both physical and psychological safety, that had not been part of the project until then. If children feel unsafe in schools, their learning will suffer.

You need to be open to everyone's opinions initially and then work out what is in the remit of the project and what is not.

7.1.6 Step 6: Watch your language

Language is key, as the above example of the field names suggests. It only takes one unfamiliar or difficult-to-understand term and people may switch off, again feeling that this discussion is 'not for them'. Check presentation material, to make sure that it is tailored to the audience. Delete all acronyms and spell them out. Even with technical stakeholders you can't assume that everyone understands the same thing by the same acronym. For example, the acronym WFD to one person could mean workforce development, but to another person it could mean the Water Framework Directive. And to members of the public, it will mean nothing at all!

As Section 7.2 suggests, it's valuable to address the effects of discrimination and disadvantage, taking care to ensure that the safe environment applies to all. This also means presenting information in a way that can be understood by all. Slides crammed with text may work for one person but not for another. Some people may understand information better if it's graphically presented, or the text is in simple bullet point form. Think carefully about how you present and disseminate information and vary the methods to meet different needs.

7.2 Equality, diversity, and inclusion

7.2.1 Whose social value matters?

We often use terms like 'community' or 'local people' as though there can be a single consistent perspective, while all communities and groups of local people are made up of individuals in all their glorious variation.

Despite good intentions, engagement often draws in only part of a community or group while other voices are not heard. This also applies to professional and organisational stakeholders. So frequently engagement events are filled with the 'usual suspects': stereotypically, these are often middle-class educated white people who are used to having their voices heard. Julie Foley, a Director at England's Environment Agency told us,

> We often hear from the voices that shout loudest in a local community: often more retired and more affluent people. The voices we are not hearing may be people who are not as willing or able to turn up to a local flood group meeting. You need to balance our community engagement to be proportionate to what is needed for a flood or coastal project, rather than saying this is the 'one size fits all' approach that we must apply to every project regardless of its scale or local risk.
>
> (Foley, 2025)

Even when others attend, they may not feel psychologically safe enough to speak up.

7.2.2 Where to focus your efforts to include everyone

We know that certain groups tend to experience discrimination and unequal treatment. In the UK, through the Equality Act 2010, it is against the law to discriminate against anyone because of:

- Age.
- Gender reassignment.
- Being married or in a civil partnership.
- Being pregnant or on maternity leave.
- Disability.
- Race including colour, nationality, ethnic, or national origin.
- Religion or belief.
- Sex.
- Sexual orientation.

These 'protected characteristics', which often reflect a history of discrimination, provide a helpful prompt to consider *is the way we are engaging likely to miss out or disadvantage particular groups*? This may include thinking about the physical accessibility of venues where events take place but also the details of how to engage, the language you use and how you treat people. It may also help you think longer term about how the project will impact different groups.

Building on this, other social and economic factors affect the stories people have to tell and also how difficult you may find it to reach them. These factors include poverty and isolation, inner city living or rurality, homelessness, and addiction. Sometimes we feel that the voices we most need to hear are absent from our engagement activities.

This matters because diversity of views is a rich and deep source of insight. A new bridge over a stream in a park might mean something very different for a commuter cutting across early in the morning on the way to the station, a young parent with a pushchair, an older person, a wheelchair user, or a jogger. Missing the range of perspectives can reduce the quality of your project, so that it doesn't work for some people – and if you miss something major, it may lead to pushback and protest during project build or after completion.

7.2.3 So, what helps?

7.2.3.1 Tip 1: Find ways to reach people other than the usual suspects

As we suggested above, go to where people are rather than expecting them to come to you: the mother and baby group, community hub, or a running club. Think about timings: a meeting at 10am may work for young parents but is not good for the commuter. A meeting at 8pm may work for the commuter but may be late for the older person. Is your venue accessible and does it feel safe for people walking home in the dark? More than one session, in person or online, is probably required, so factor them into your engagement plan.

7.2.3.2 Tip 2: Think about what matters to people

This helps how you frame your engagement. You can position the imaginary new bridge project in a number of ways:

- Protecting biodiversity – by reducing the footfall along the side of the stream.
- Flood mitigation – so when the stream floods there's a way through.
- Improving public access – so more people can use the park and get to the station.
- Modernisation – upgrading the local environment for the future.

These different ways of positioning the discussion may matter more or less to different members of the community.

You may want to think about social value at a higher level. In the UK the 'Britain talks climate' project (Wang, Corner, and Nicholls, 2020) used a range of psychological and ideological factors, including threat perceptions, social identity, and belief in meritocracy to define how seven different segments of society engaged with the issue of climate change.

For instance, what they call 'civic pragmatists'

> try to follow a low carbon lifestyle, but feel demotivated by a lack of political ambition on climate change and other social issues. Reflecting their pragmatic nature, they are likely to look past their opinion of the government of the day and support progressive climate policies when they see them.

On the other hand, 'Backbone conservatives' 'care deeply about food, farming and the rural economy. But they are more sceptical about grand claims of global leadership, or the "virtue signalling" of (what they sometimes see as) symbolic lifestyle changes' (Wang, Corner, and Nicholls, 2020 p.7).

Similarly organisational stakeholders have different cultures and priorities. We worked on a project about prison education which brought together staff members from the prison and probation services, colleges, and education providers. Not surprisingly, and reassuringly for most of us, the culture of the prison service was risk averse and focused on avoiding potential problems. The college team were used to thinking about people's potential and opportunities for expansion. When we began to explore the scope for online learning, the prison staff team saw the issue through a lens of risks and threats, while the college team had a lens of learning and growth. It was very difficult to find common ground.

The following case study describes a school project which puts social value at the centre of engagement.

Case study: Westpark Middle School, Red Deer, Alberta, Canada

Westpark Middle School in Alberta, Canada, was constructed in 1968. In 2015, Berry Architecture + Associates began an investigation to determine the condition of the existing school building and whether there was a need for a new school and if so, what important features could be included.

In 2016, Berry Architecture won the contract to complete the design for a new Westpark Middle School. A fundamental principle of Berry Architecture's approach was a commitment to open stakeholder engagement and working with the students to include their social values in the design of the school. The school has a diverse mix of students from a variety of cultures and backgrounds, including First Nations and a history of supporting many students with learning challenges and physical mobility issues.

In spring 2016, as part of the early schematic design phase of the project, the Berry Architecture team members began to meet with the teachers and students. These discussions involved current students, as well as future students attending the school as young as ten years old. The intent of these discussions was to determine what was important to the students.

Social value was a theme that the design team built on. When talking with the children, they were asked about what they wanted in a new school. There was an expectation from the design team that they would receive responses like: a skateboard party, video arcade room, and music systems and TVs everywhere. Instead, the design team received ideas like an open and easy entry for students in wheelchairs, a visible elevator so students with mobility issues didn't have to go around to a back corner to access it and a variety of spaces for students to get together with no hidden corners to discourage private conversations and exclude others. The students spoke a great deal about inclusion and not wanting to create a school that would create areas for small groups to develop and exclude others. They were concerned with features that would improve the quality of education for everyone.

The engagement process involved a series of planned meetings where the design team taught the student design team about various aspects of architecture and planning. The students were encouraged to think big and not focus on budgets. Their focus was always on doing what was right for all the students, staff, and visitors to the school. Whether the students knew it or not, their social values strongly influenced them, the design process, and ultimately the final school project.

Berry Architecture had students from the school design group come spend a day in their office. During this time, all the project consultants (in-house interior designer, mechanical engineer, structural engineer, electrical engineer, civil engineer) attended the day and gave a brief presentation about their profession and facilitated some smaller group discussions. This provided the students with an insight into what the profession does and what a career could look like.

Conducting a design process in such a way was very rewarding for the Westpark Middle School design, Red Deer Public Schools, the local Westpark community and all the consultants involved. Westpark Middle School quickly developed a reputation for being a caring and inclusive school with the values of the students and community visible through the school design (Figure 7.1).

Figure 7.1 A design meeting with students.

Source: Berry Architecture (2025).

Lessons from the stakeholder perspective

- The design team worked at participant's level whether they were students, teachers, school elected officials, or the general public. They tried not to talk above their knowledge level and never used unexplained jargon.
- The design team provided feedback after the various engagement sessions to those who wanted it. This demonstrated that what was being said was heard, considered, and implemented where possible.
- It is important to be open-minded, listen to all perspectives and stories related to the project and consider all of the ideas presented and be respectful of those involved.

A former associate superintendent from Red Deer Public Schools indicated that the engagement process used for Westpark Middle School was the most insightful, in-depth, and considered for a new school project that they had seen.

Lessons from the social value perspective

- Inclusion is for everyone, regardless of physical or mental ability.
- You can't assume that you know what people want – you need to ask them and listen, and you will probably be surprised at what you hear!
- Stand up for what is right and don't question your values.

Takeaway

The engagement process that Berry Architecture used for the new Westpark Middle School was an example of what can happen when designers listen to

Figure 7.2 A design team meeting with student representatives.

Source: Berry Architecture (2025)

the end users. Design features were included that may not have been initially thought about, but by understanding the students and users of the school's social values, a design was produced that truly represented those who were to use the school and those future generations to come (Figure 7.2).

7.2.3.3 Tip 3: Vary how you engage

It follows that you would use different messages and arguments to engage with each of the segments of society' around climate change in general, and the technical aspects of any particular project. This requires flexibility to adapt messaging according to the audience.

The same applies to channels. Different groups are familiar with different ways of communicating and some may be missed if you communicate through unfamiliar channels. Most professionals still use email in 2025, while many members of the public may be more used to messaging and group discussion apps. When we worked with a group of members of the public, we duplicated all communication through email and text.

The range and variety of perspectives and issues to look out for can seem overwhelming. The key is to consider how your engagement – methods, messages, and style – will land with different groups of people. A variety of methods will help your engagement land well and make a meaningful response more likely. Again, this should be reflected in the engagement plan.

7.3 Re-focus engagement

The traditional approach to engagement is starting to lose its hold on the built environment. Over the years we have seen a shift away from traditional methods in which stakeholders and communities are told about projects, and an increase in project teams wanting to consult, collaborate, and co-produce schemes with stakeholders.

An example is the UK Environment Agency's approach, developed over time through advice and practice (Environment Agency, 2009). In the past the DAD (Decide, Announce, Defend) approach was employed on flood risk management schemes. This involved deciding the best course of action based on professional assumptions using processes and metrics such as cost-benefit analysis. The next step was to 'announce' or tell affected stakeholders what was being planned, what it looked like, why this was the best approach and what would happen. Technical experts would run sessions with communities and stakeholders to present maps and visuals of what the scheme would entail. This was very much an information sharing exercise, with little room for input from stakeholders as decisions had largely been taken.

This brings us onto the final stage: 'defend'. The announcement was often the first time many stakeholders had heard about the project. The near complete nature of the project meant that if there was opposition, defending the design was the only option. Time, money, and resources had been spent in developing the design to this stage. There was no money for redesign should the stakeholders require it. Similarly, presenting a near complete design meant that stakeholders felt powerless to do anything but accept it … or oppose it if their feelings were strong enough.

This approach was fraught with problems. Tokenistic engagement with little or no chance for stakeholders to have a say created a tense and unproductive relationship between the Environment Agency and stakeholders on a lot of projects.

Like other practitioners in the built environment, the Environment Agency no longer recommends the DAD approach. Instead, they advocate EDD (Environment Agency, 2009). EDD shifts the focus away from professional opinions and perspectives to a much more collaborative approach.

The first stage is to engage. Before a design is anywhere near complete the project team is encouraged to engage with stakeholders to understand the situation and share their thoughts, opinions, and experiences. Gathering information at this stage means that it can be incorporated into the design. It reduces the need to make changes later in the project and in principle saves time and costs across the project lifetime.

The next stage is for the project team and relevant stakeholders to deliberate over what to take forward to the next stage of design. Managing expectations is a critical part of the EDD approach as both the project team and stakeholders need to understand how much influence stakeholders will have. Explaining clearly what can be influenced by stakeholders is important to ensure that they feel that their voices are heard without raising unrealistic expectations which could lead to conflict later.

The final stage of this approach is for the project team to decide how to take the design forward. This will have been facilitated by discussions with stakeholders, so opportunities and constraints should be apparent. Stakeholders have had the opportunity to feed in suggestions to improve the project, and to avoid negative effects. Of course, there may still be defending to do, but the project team are working on a much more secure basis.

The EDD approach minimises shocks to stakeholders and consequently should reduce opposition as the project develops. EDD helps project teams understand what stakeholders want and need, but also where any contentious issues could be. This helps design out issues which could cause opposition, positively impacting the project's budget and programme.

7.4 How facilitation can help

To enhance the process of engagement, you may think of calling on a facilitator to support some of your conversations with stakeholders.

A facilitator is a person who helps the organiser and participants in engagement events, particularly meetings with stakeholders, achieve their objectives. They are not a subject matter expert, indeed a professional facilitator brought in to support a project may know very little about the subject matter. They focus on creating a process to enable engagement objectives to be achieved while acting as an independent, neutral party.

In engagement events the facilitator works:

- To create and sustain a participatory environment, enabling all participants to contribute on equal terms.
- To handle conflict and disagreements that may emerge.
- To guide the group to appropriate and useful outcomes (International Association of Facilitators, 2021).

Having a facilitator takes the pressure off the project team who no longer have to think about handling meeting dynamics, time keeping, or what to do about disruption. The project team can focus on the content of the meeting including being able to contribute to the discussion. An engagement professional may have facilitation skills and use those as part of a project team. In that case, someone else in the team should take responsibility for presenting and responding on behalf of the team. Alternatively, you may bring in an independent facilitator from outside the team.

On a practical level a facilitator will help keep the discussion on time and on track, and think about how the group can best achieve objectives including considering when to break into smaller groups or use tools like meeting polls.

Because the facilitator has no 'skin in the game' they create an environment where everyone can contribute on equal terms and ensure that all contributions are valued. Even with the best of intentions, project team members may unconsciously steer the meeting towards their preferred conclusion (Harrin, 2020). The facilitator is not tempted by that, so having a facilitator can be particularly helpful when working with multiple stakeholders or members of the public.

7.4.1 When is a facilitator most valuable?

Most meetings would benefit from expert facilitation, but you should consider it particularly in the following contexts:

- Where neutrality and independence are important, for instance in discussing contentious issues with the communities.
- Where multiple stakeholders are involved and your team wishes to participate as an equal party rather than being seen as the lead player.
- Where you expect conflict, challenge, or contentious issues to arise with the potential for raised temperatures.
- Where the engagement is complex, has a lot of moving parts, and/or large numbers are involved.

7.4.2 What do facilitators do?

Facilitators typically work with the project team to develop the agenda and programme for a meeting or series of engagement events: what facilitators call 'the process'. They help the organisers confirm the objectives they are seeking to achieve. This includes thinking about things like the balance of presentations and discussion, when to break into smaller groups, and when to use tools like meeting polls or sticky notes.

In the session itself the facilitator manages the process, enabling the project team to focus on the content of the meeting. This includes handling issues such as challenge and conflict, ensuring that everyone can participate, and drawing the meeting to a close on time with the objectives achieved!

7.4.3 Key facilitation skills

In Chapter 4 we summarise stakeholder engagement skills required to support socially successful projects. Facilitation is one of those skills. Looking in more depth we would summarise key facilitation skills and activities as follows:

- Designing an agenda, session plan, or series of engagement activities based on objectives.
- Designing processes that encourage inclusive participation by all members of a group.
- Understanding and working with group dynamics.
- Listening, questioning, and summarising discussions.
- Using collaborative engagement tools both online and in person.
- Handling issues that arise such as conflict, challenge, dominant, and quiet participants.
- Reflective practice and continuous learning (International Association of Facilitators, 2021).

7.5 How does it all hang together?

So far, we've described some of the ingredients of socially successful projects, and in particular the linked ideas of social value and stakeholder engagement. How do you move from a kitchen full of 'stuff' to a satisfying dish? What's the menu? How messy is the cooking experience?

7.5.1 Complexities and pitfalls

Sometimes you have all the ingredients, follow the recipe and the result just isn't right. You may need to try again and vary amounts, seasoning, cooking time – even improvise a new approach to get the result you want.

We learn how to create socially successful projects by trial and error and learning from experience. And we recognise that no two projects are the same.

Getting this right involves some simple principles but it is also inherently complex, not just because different people and groups have different social values, important as that is.

7.5.1.1 We may not understand what really matters to them

We may ask the right questions about what people value, but get back a superficial or partial response which we may not fully understand. When the schoolkids were talking about the toilets they were really talking about safety, but it took time and some probing to understand that. People's responses may be unclear and may need interpretation and enquiry to be understood

7.5.1.2 We bring our own biases and assumptions

Linked to this, we bring ourselves into the discussion and we all have biases, assumptions, and heuristics (mental shortcuts) some of which are useful and some less so.

We may make assumptions about what people want. In one project we assumed that older people would rather meet face to face rather than online (an assumption about technical ability) whereas they told us that they were happy with online meetings as they didn't like to travel in the evening (for them safety and convenience were more important).

We may make assumptions about our projects. In one project, the project manager produced photos of two possible embankment structures, one all concrete, the other a grassy sloping bank. She assumed that everyone would prefer the sloping bank (because she did) and was astonished when some members of the community preferred the concrete (which was stark and rather grand in their eyes).

7.5.1.3 Where social success fits

And that's before you even get to the relationship between the technical, financial, and social aspects of your project. There are usually trade-offs – for instance the community want an embankment to feature a wide 'boardwalk' for cycling and scooters but this is pricey and technically hard to achieve. So, what gives? Julie Foley, Director of Flood Risk Strategy and National Adaptation at the UK's Environment Agency told us 'You need to weight your community engagement to what is needed for the project, rather than saying this is the approach that you must apply to every project' (Foley, 2025). How do you discuss trade-offs relating to values which are by their nature important to people? It's important to make clear from the outset what is open for discussion and what's not, sharing the details of budget, or some of the main technical restrictions to manage expectations.

7.6 Successful early engagement

Getting the early engagement recipe right brings multiple benefits.

7.6.1 Save time and money

Listening to stakeholders from the early stages of the project to understand what matters to them challenges project teams:

- To think beyond 'the usual suspects' to engage with the stakeholders whose voices really count.
- To sharpen up engagement materials and messaging.
- To involve stakeholders early and listen more carefully.

… all of which takes more time! But the aim is to use time and resources more effectively and above all reduce the costs that come with poor quality or misdirected engagement. The aim is to get engagement right the first time.

7.6.2 Reduce stress

Demonstrating that you understand and value the contributions of stakeholders and communities leads to an easier path even for the most contentious projects. The social dimension of projects can be a cause of stress in particular if you speak to the wrong people using the wrong language. You then have to manage reactions ranging from anger to cynicism. If you miss out groups early on and have to bring them in later, catching up can be difficult.

Getting it right the first time reduces stress for both you and the stakeholders, contributing to psychological safety on both sides and ultimately building greater trust.

7.6.3 Enhance your reputation

People really care about how they are engaged. Poor quality engagement – late, inappropriate, failing to listen – makes people angry and they are likely to tell others about their experience. Getting it right the first time can build your reputation. Good engagement feeds on itself, as the word gets around that you mean it and that people will really be listened to.

7.6.4 Better projects

Listening to understand the social values of stakeholders and communities always brings insight. It may challenge your thinking about what really matters about a project. Project teams may need to shift their mindset from seeing engagement as a necessary evil, something they have to get out of the way, to a valued source of knowledge. Challenge is healthy for every project, wherever it comes from. In our experience, deep engagement with stakeholders' social value often makes for better project results even in technical terms as you gain new information and new ideas are suggested. Beyond that, knowing that your project embodies what matters to people is a great result.

7.6.5 Insights for future projects

Working across multiple projects over the years we have seen teams make the same mistakes as they wait too long to engage, fail to listen, or fail to act on what they hear. It's easy to waste time and energy. Really listening to stakeholders also provides great learning to take forward to future projects. Like any successful recipe, you can perfect it over time.

Key takeaways

Creating psychological safety is vital to engagement and requires time and resources.

Find ways to reach beyond the 'usual suspects' for your engagement. Really listen and show that you are listening.

Expert facilitation can help create and sustain a participatory environment and relieve pressure on project teams.

Engage early and keep at it.

Understanding what matters to the full range of stakeholders can benefit your project, reduce stress for the project team, and provide learning for the future.

References

Babaei, A., Locatelli, G. and Sainati, T. (2023) Local community engagement as a practice: An investigation of local community engagement issues and their impact on transport megaprojects' social value. *International Journal of Managing Projects in Business* 16:3, pp. 448–474. https://doi.org/10.1108/IJMPB-10-2022-0224

Edmondson, A. C. (1999) Psychological safety and learning behaviour in work teams. *Administrative Science Quarterly* 44:2, pp. 350–383. https://doi.org/10.2307/2666999

Environment Agency (2009) *Mainstreaming collaboration with communities and stakeholders for FCERM Improving Institutional and Social Responses to Flooding Science Report: SC060019 Work Package 4* [Online]. Available at: https://assets.publishing.service.gov.uk/media/602e6d838fa8f5432d802fe6/mainstreaming_collaboration_with_communities_and_stakeholders_for_FCERM_WP4_technical_report.pdf. Accessed 23/08/2025.

Fischer, F. (2000) *Citizens, Experts and the Environment: The Politics of Local Knowledge.* Duke University Press: Durham. https://doi.org/10.2307/j.ctv11smwd9

Foley, J. (2025) 'Book Interview'. Interview by Simon Wilson [Microsoft teams], unpublished, 30/07/2025.

Harrin, E. (2020) *Engaging Stakeholders on Projects.* Association for Project Management: London. Accessed 14/7/2025.

International Association of Facilitators *Core competencies*, 2021 [online] Available at: https://www.iaf-world.org/wp-content/uploads/2025/08/IAF-Core-Competencies.pdf. Accessed 14/7/2025.

Newman, A., Donohue, R. and Nathan, E. (2017) Psychological safety: A systematic review of the literature. *Human Resource Management Review* 27:3, pp. 521–535. https://doi.org/10.1016/j.hrmr.2017.01.001

Queen, C. (2025) 'Book Interview'. Interview with Sarah Fitton and Simon Wilson, [Microsoft teams], unpublished, 20/02/2025.

Remen, R. N. (2006) *Kitchen Table Wisdom: Stories That Heal,* 10th Anniversary Edition. Riverhead Books: New York.

Scott, J. (2001) *Power*. Blackwell: Oxford.

Wang, S., Corner, A. and Nicholls, J. (2020) *Britain Talks Climate: A toolkit for engaging the British public on climate change*. Climate Outreach: Abingdon.

8 Situational analysis and scenarios

> It ain't what you don't know that gets you into trouble. It's what you know for sure that just ain't so.
>
> (Attributed to Mark Twain)

Key steps

- Scan the horizon and analyse the context.
- Create scenarios to explore.
- Involve others in developing scenarios.

Context is key.

Before diving into stakeholder engagement planning, it's important to understand the context and environment. This helps situate what you are planning. It challenges you and others to think beyond the immediate. It helps you understand the context in which people's social value is developing.

Imagine a government department setting off to develop a new policy with major implications for society. Now imagine that a general election is due in a month's time. Probably this is not the right time to press on. It may be wise to delay or at least scale back the preparations until the context becomes clearer.

Usually what's shown by the horizon scan, or situational analysis, is less evident than that example. It may reveal developments and trends that require thought about how to position engagement.

For a project to build a water treatment plant, for example, this analysis might cover wider development plans for the area affecting demand; potential legal changes or challenges that could affect what is permitted; whether businesses will invest because of benefits from the plant; local politics and community agendas that might favour or oppose the idea of the plant. Thinking early about these factors helps avoid nasty surprises further down the track and encourages flexible

DOI: 10.1201/9781003659877-12

planning. It may also help identify key stakeholders with whom to build relationships early on, who may be able to tell you things you do not already know.

In this chapter we look at how horizon scanning and situational analysis can enhance stakeholder engagement.

8.1 Horizon scanning and situational analysis

Horizon scanning helps you spot 'signals': trends or developments that may not be vital now but which may become influential over time. Twenty years ago, concern about climate change was a 'signal', something to bear in mind. For the water treatment plant, it might be a rumbling concern in the local community about a planned shopping centre, or local authority plans to change planning policies. These are things to keep an eye on and factor into the 'what if' aspects of your planning (Sutherland and Woodroof, 2009).

Horizon scanning prompts questions about a wider range of issues than you might otherwise consider, particularly if done systematically over time.

This stage of the project requires a combination of some tools for analysis and, perhaps even more importantly, ways of thinking including systems thinking, anticipatory thinking and critical thinking enabling you 'to navigate uncertain futures and to act better and more proactively towards achieving preferred futures, or to be better prepared to face (or avoid) undesirable ones' (European Environment Agency, 2023, p.8).

The tools and methods in this chapter can all be used early in a project. Typically, they would be considered in a workshop or meeting of project team members. We recommend that you also involve stakeholders in these discussions from an early stage to bring wider perspectives and test your thinking. It is also worth redeploying these tools and methods at different project stages for longer projects, for instance as you move from design to construction and operation.

Let's look at some of these tools now.

8.1.1 Tool 1: PESTLE

A popular set of prompts started life as STEP – encouraging people to think about potential developments in **s**ocial, **t**echnological, **e**conomic, and **p**olitical domains. This was later expanded to PESTLE (or PESTEL): **p**olitical, **e**conomic, **s**ocial, **t**echnological, **l**egal, and **e**nvironmental domains (Yüksel, 2012).

You can use these prompts in a workshop or planning session to ask participants to consider likely developments under each of these headings, building up a picture of the future 'horizon' that may form the context for the issue you are discussing. This is commonly done with the project team, using their technical knowledge to inform the assessment. However, involving stakeholders with whom you already have a good relationship with will bring an added valuable dimension.

Some examples of what might be generated for the water treatment plant case are set out in Table 8.1.

Table 8.1 The PESTLE methodology applied to the case of a water treatment plant. Based on Yüksel (2012)

Area of focus	*Future considerations*
Political	Potential change of local governing and/or national governing party Growth of populism Climate scepticism and climate awareness
Economic	Cost of living concerns Tight funding for public projects Growth (or decline) in employment projected in the area
Social	How outdoor spaces are used post pandemic Polarisation of views and reflection in social media Demographic changes in the local area Trends in crime and anti-social behaviour
Technological	Emerging technologies for plant design and construction Citizen access to data via shared technologies
Legal	Regulatory planning framework
Environmental	Assessment of local biodiversity Flood and drought risks including impact of climate change Public awareness of environmental concerns

Using PESTLE for horizon scanning helps shake you out of the familiar grooves and ways of thinking. For the water treatment plant, thinking about technology and social media might prompt ideas about how and how far to share data from the plant operations with the public. Thinking about crime and anti-social behaviour might affect the security aspects of the design such as fencing and lighting.

Ideas from the PESTLE analysis should be logged and may feed into different elements of the project plan, including technical design, budget planning, and engagement planning.

8.1.2 Tool 2: Situational analysis

An alternative, or an addition, to horizon scanning is a checklist approach listing key factors in the situation for your project.

An example is the following list, based on a checklist developed by the Environment Agency in the UK. In Table 8.2, we have populated the list with examples of the factors you might identify for the water treatment plant.

Again, this might be used in a planning session with the project team, with other stakeholders involved if possible. They would work through the checklist to identify issues that might arise with the project. Some of this may be already available information whereas some information may need to be developed through engagement during the project. One report on the planning system referred to these as 'passive data' and 'active data' (Mitchell et al., 2022). Again, issues identified would be fed into the project plan, for instance allowing time to address regulatory and consenting procedures.

Table 8.2 Situational analysis example for the water treatment site case. Based on Environment Agency checklist

Factors for consideration	*Examples*
History of engagement	Previous discussions with community groups have revealed strong interest in preserving heritage sites. Local authority engagement has been volatile because of frequent changes in political leadership.
Physical considerations	Previous ground surveys have identified utility cables and redundant underground storage tanks from the 19th century. Responsibility for the storage tanks is unclear.
Cultural, heritage, archaeology, and material assets	Park site includes listed ruins, a river bridge, and working cafe site.
Timing	Major redevelopment of an adjacent commercial site will be undertaken by the local authority next July. It will be important to align the two projects and potentially engage stakeholders around both projects at the same time.
Estates, access	Access on the North side is via a private road owned by a community trust who will need to grant permission.
Decision-makers	Decision-makers include the local authority, the community trust, and the planning authority.
Local plans, legalities, and policies	In addition to the planning regulations, the project will need to consider existing Tree Preservation Orders, national requirements to achieve biodiversity net gain, and eventually health and safety regulations for all work to be carried out, and highways regulations for access.
Assessment/consenting/ permitting regime	These will include potentially scheduled monument consents, and tree works permitting.

In terms of layers, horizon scanning with a tool like PESTLE gives you a high-level context whereas the situational analysis brings you closer to practical implications for a local or sectoral context. Both provide valuable insights.

8.2 Scenarios

The next step is to build on horizon scanning and/or situational analysis and begin to construct scenarios.

This helps to overcome the common risk of believing that there is only one future (or one scenario) for your project with a linear track to reaching it. In practice the vision and process you begin with is invariably not the one you finish up with. Instead, it is more useful to think about a range of possible futures or scenarios for your project. One practitioner categorised the different types of scenarios under different headings, including 'preposterous, possible, plausible, probable and preferable' to cover most eventualities (Voros, 2017).

Scenario planning consists of using a few contrasting scenarios to explore the uncertainty surrounding the future consequences of a decision. For conservation as an example, the major benefits of using scenario planning include:

1 Increased understanding of uncertainties.
2 Incorporation of alternative perspectives into planning and design.
3 Greater resilience to decisions which are unexpected (Peterson, Cumming, and Carpenter, 2003).

Here we focus on scenarios that relate to stakeholders and engagement. The technique has much wider application, and it may be more useful to develop scenarios for the project as a whole and then consider implications for engagement.

Usually, it helps to choose scenarios that are plausible and internally consistent, creative, and relevant to your project, the project team, and your stakeholders (Amer, Daim, and Jetter, 2013). If we look at the water treatment plant example, you might construct three scenarios for the project:

Scenario 1: Strong support for the development of the plant from the local community and planning authorities.

Scenario 2: Private sector interest in funding which would require changes to design and location.

Scenario 3: Local opposition becomes intense with a media campaign against the supposed environmental and amenity concerns of the project.

Maybe also consider something more preposterous or possible to stretch your thinking: perhaps a game-changing technology, or changes in water use that could shake the foundations of the project.

The next step is to consider the implications of each scenario for your project. This could be a discussion among the project team, ideally with stakeholders. The purpose of scenario planning is not to predict the future, but to work out what needs to be in your project and engagement plans to build resilience for whatever the future brings. Key questions at this stage for each scenario could be:

- What are the upsides and downsides of this scenario?
- Who would be involved?
- What does our planning need to include so we are resilient in the case of this scenario?

The prime deliverable from the work with scenarios is measures to strengthen the resilience of your project and engagement plans. Having considered scenarios at the early stages of a project, it is useful to revisit assumptions later on, exploring what may have changed and what new information may be telling you about potential scenarios, and what will make your project, and engagement plans more resilient.

8.3 VUCA

Another framework that can help with identifying and mitigating certain situations is the VUCA framework (**V**olatility, **U**ncertainty, **C**omplexity, and **A**mbiguity). Like many concepts in business management, the acronym VUCA came from the military and was originally used to describe the world after the end of the Cold War. It was then taken into civilian use, and many organisations are now aware of

Table 8.3 VUCA as it applies to a water treatment plant. Based on Bennis and Nanus (1985)

Situation	*Mitigation*
Volatility: Political uncertainty and potential changes in local leadership at short notice.	Keep on your toes. Think about which issues and which stakeholders might be ready to 'blow'. Have contacts with all political groups and be ready to flex messaging to address political concerns.
Uncertainty: Population growth will have an impact on demand for plant use. But future population growth is uncertain because of planning constraints.	Have backup plans in mind with a plan B in case planning enables major growth in the local population. Don't try to predict but go with the flow.
Complexity: There are multiple communities locally all with different perspectives, ranging from environmental pressure groups to those committed to 'heritage protection'.	Resist the lure of simple solutions. Think about who can help you understand the interconnections – who has the knowledge and insight? Be open to it not necessarily being an industry professional so cultivate local contacts who can provide insights.
Ambiguity: Anticipated public health improvements may be attributed to better sanitation or public awareness campaigns. Past case studies do not serve as reliable guides. Unforeseen opposition, for example, protests against plant location due to odour concerns.	Be ready to be surprised and embrace the unknown which may include emerging new data sources, relationships, and alliances. There's a first time for everything. Be prepared to improvise.

the idea that our projects may feature some or all aspects of volatility, uncertainty, complexity, and ambiguity (Bennis and Nanus, 1985).

We see VUCA as less a tool and more a way of thinking about the nature of the issues you are facing, and how you might mitigate them. Nowhere does this matter more than in how we understand stakeholders and the behaviours we observe.

Table 8.3 sets out the VUCA factors, how they might apply to the water treatment project, and how the team might respond effectively.

8.4 Involving stakeholders in horizon planning and scenario development

Horizon scanning and scenario planning are creative processes built on qualitative data. It follows that bringing diverse perspectives to the table enhances the reach and flex of this stage. Stakeholder voices make thinking more robust, and they may surface further trends and scenarios.

Cognitive biases affect how we make judgements and decisions. For instance, confirmation bias leads us to pay more attention to the views and evidence that confirm our beliefs, and anchoring bias leads us to pay more attention to the first pieces of information we encounter (European Environment Agency, 2023). Diverse stakeholder voices can provide salutary challenges to those biases early in a project.

Truly effective work at this stage can challenge thinking and change minds. In his highly influential (and still very readable) article about developing scenario planning for the oil industry, Pierre Wack (1985) wrote: 'We now wanted to design scenarios so that managers would question their own model of reality and change it when necessary, so as to come up with strategic insights beyond their minds' previous reach' (Wack, 1985, p.85).

Since that time, scenario planning has become widely, if inconsistently, used, in planning and development. 'At its core, scenario planning guides planners, community members, and other stakeholders to consider the various futures they may face – good, bad, and unexpected' (Goodspeed, 2019, p.1). It is helpful at this stage to challenge team members to think differently and to build the flexibility and resilience to tackle whatever the future may bring.

However, there are practical implications. Deciding who to involve at this stage, and the time commitment required, are key considerations. It can seem counterintuitive to bring stakeholders together to discuss a big picture topic like 'what would the future look like'? Yet for major projects this type of engagement builds understanding of the issues and involves stakeholders in creating a collective understanding of the future. It helps build relationships with stakeholders that will be useful in later stages.

The following case study shows how involving citizens in big-picture thinking can be beneficial.

Case study: Citizens panels

The concept of 'adaptation pathways' has been developed as a way of thinking about long-term planning for complex challenges such as climate change and flood risk (Environment Agency, 2025a). The idea is that, rather than developing a single plan, a pathway can be elaborated with 'forks in the path' at different points, enabling a flexible response to be implemented depending on the speed and impact of climate change (Haasnoot et al., 2012).

We worked on a project exploring how the concept of adaptation pathways could enable planning for future flood risk in the context of climate change (Environment Agency, 2025b).

The citizens' panels brought together broadly representative groups of citizens from four geographical areas. The future of each area was threatened with long-term flood risk with resulting challenges for housing development, farming, and business, and how we live in the future.

Engagement objectives were summarised as follows:

- To understand what matters to residents in relation to flood risk.
- How to build trust and collaborative partnerships.
- Understand challenges, solutions, and opportunities in the context of adaptation pathways.
- How to include residents in decision-making.

The aim was to engage a broadly representative sample of people from the different areas. This involved balancing the sample group so that factors such as race, class, and age aligned with the local population. This was achieved through 'sortition': recruiting a pool of around 100 people at random and then sorting for different characteristics.

Delivery involved facilitating panel discussions featuring external speakers during an 'educate' stage, use of scenarios and personas to stimulate thinking about options and alternatives during a second stage, and finally agreement on recommendations by panel members.

For this project we developed a series of scenarios and asked the panel members to explore the implications. One version took the form of 'personas' where the scenario was personified, that is, described in terms of an individual with their wants and needs who would be affected by climate change.

In facilitated workshop sessions, the panel members discussed how things might play out in the different scenarios. They did not choose a preference, or solve the problem, but created a broader understanding of the complexity and interrelationship of issues involving many stakeholders. Learning from the panels was taken forward by the relevant project teams for later stages of development.

The project led to changes in the approach to adaptation pathway planning in local areas, including considering how to involve the public in future developments and focus on social value. Projects aimed to continue the conversation, particularly through informal channels, leveraging traditional and digital media.

The project registered a significant impact on the panel members themselves who felt informed and empowered. At least one panel member was considering a career change to become more involved in environmental issues.

Evaluation anticipated longer term impacts in the approach of adaptation pathway and other projects from the involvement of the public. It found that panels help to reach diverse groups who might not normally participate in consultations on complex issues.

Overall, the main influence on decision-making was understanding that ordinary citizens can engage with complex concepts such as adaptation pathways and develop balanced recommendations about how they should be taken forward.

Lessons from a stakeholder engagement perspective

- Engaging a broadly representative sample of a population can lead to significant insights into complex issues with long-term implications.
- This is best supported by an educational component presented by local or national experts in different topics.
- Scenarios and personas helped engage citizens around complex issues.

Lessons from a social value perspective

- Use of scenarios and personas helped draw out value-based perspectives from participants including the emotional component of the issues.
- The deliberative nature of the engagement helped participants look beyond short-term interests to consider the range of options that might benefit larger populations over time.

Takeaway

Involving citizens in deliberations on complex, long-term issues can reveal insights and recommendations to support policymaking.

8.5 Situational analysis and scenarios for socially successful projects

A social value lens helps re-focus situational analysis and scenarios to give equal weight to social, technical, and financial factors.

8.5.1 Broadening horizons

Over the years we've facilitated many horizon scanning sessions and unsurprisingly participants tend to focus on the factors and drivers they know well. With a group of engineers, the flip chart for 'technological' factors fills up fast, while the 'economic' chart is full with a group of economists.

A social value lens points towards equal weight for the economic, social, and environmental aspects of the scan, and of course the other PESTLE factors are equally important.

What helps a group broaden their horizons during this stage? We recommend four things:

1 Provide prompts to thinking in advance, encouraging team members to read in unfamiliar subjects and stretch their thinking.
2 Involve a diverse group, potentially including stakeholders from outside the team, to bring different perspectives.

3 Iterate, that is go over the horizon points several times to spark new thinking.
4 Effective facilitation: use a dedicated facilitator who is not part of the team prompting, and where necessary, challenging the group to look and think more widely (see Chapter 7).

8.5.2 Whose scenarios are they?

A lot has been written about how to construct a good scenario (e.g. Voros, 2017; Wack, 1985), and it's worth spending time to make sure they are fit for purpose. That includes being acceptable to the stakeholders who you want to involve in the discussion.

A key principle for us is involving stakeholders early, so why not bring in different views at this stage to help shape thinking about context? This also exposes a wider group to the challenges and constraints in the external environment. This understanding is helpful if further down the track you are presenting options that could be unpopular with them.

8.5.3 What type of scenario?

Scenarios need to strike the balance between conveying risk while not being frightening.

Using PESTLE as a series of prompts is effective, provided it's not mistaken for a prediction tool. The risk is that it offers a 'siloed' view and neglects the overlaps. So, a 'sustainability scenario' for the water treatment plant might be: *'Local people appreciate how the plant will bring new jobs to the community and improve public health, with a biodiversity action plan to enhance the environment'.*

The three 'traditional' scenarios above are used to test the robustness of your plan. The sustainability scenario is more like a first step towards a sustainable vision for a socially successful project.

Again, involve stakeholders in both developing and exploring scenarios so they understand the tensions and trade-offs involved in making your project socially successful. You are also showing them the constraints and parameters you have to work with and increasing their understanding of your perspective. Social value is about challenging what you know and becoming open to new thoughts and perspectives, helping you reduce the risks around a project and potentially finding new solutions.

8.5.4 How understanding social values contributes to horizon scanning

Involving stakeholders and teasing out their values offers a unique contribution to this stage of projects.

Looking more deeply at the trends that affect our projects, we can think about three stages (Spencer and Montero Salvatico, 2019):

- Point of manifestation: what we can observe by scanning from an external perspective.

- Point of origin: shifts in social values that may not have worked through into day-to-day conversations.
- Point of impact: where a trend becomes apparent and we can work on the implications of that trend.

Engaging in dialogue with stakeholders and communities can help you literally get ahead of the trend, identifying what really matters not just for the present but also for the future. This is extremely important given the inter- and intra-generational nature of built environment projects.

It will also provide a base for the next stage of work: exploring which stakeholders will be critical to the success of the project, which we look at in more detail in the next chapter.

Key takeaways

Take the time to understand the different situations that you and the project could be faced with.

Understand the context and involve stakeholders in understanding different scenarios.

Be open to the new thinking, ideas, and values that stakeholders bring.

Remember that you are not predicting the future, but building sufficient flexibility to meet what the future brings. This will help to plan mitigation measures if necessary or just not to be caught off guard.

References

Amer, M., Daim, T. U. and Jetter, A. (2013) A review of scenario planning. *Futures* 46, pp. 23–40. https://doi.org/10.1016/j.futures.2012.10.003

Bennis, W. and Nanus, B. (1985) *Leaders: Strategies for Taking Charge.* Harper and Row: New York. https://doi.org/10.1002/hrm.3930240409

Environment Agency (2025a) *Adaptation pathway programme* [Online] Available at: https://engageenvironmentagency.uk.engagementhq.com/adaptation-pathway-programme. Accessed 13/06/2025.

Environment Agency (2025b) *Communicating climate change information for flooding and coastal erosion* [online] Available at: https://assets.publishing.service.gov.uk/media/67dc3e59cb8c6838d74b5000/FRS23249_Communicating_climate_change_information_for_flooding_and_coastal_erosion_R1.pdf pp 59–60 Accessed 13/ 07/2025.

European Environment Agency (2023) *Horizon scanning — tips and tricks*. European Environment Agency: Copenhagen.

Goodspeed, R. (2019) *Scenario Planning: Embracing Uncertainty to Make Better Decisions.* Lincoln Institute of Land Policy: Cambridge.

Haasnoot, M., Middelkoop, H., Offermans, A., Beek, E. V. and Deursen, W. P. V. (2012) Exploring pathways for sustainable water management in river deltas in a changing environment. *Climatic Change* 115:3, pp. 795–819.

Mitchell, K., Hughes, J., Mulholland, C. and Samuel, F. (2022) *Delivering better place outcomes through use of data: building on the emerging digital planning system.* Research Report, Stantec.

Peterson, G. D., Cumming, G. S. and Carpenter, S. R. (2003) Scenario planning: A tool for conservation in an uncertain world. *Conservation Biology* 17:2, pp. 358–366. https://doi.org/10.1046/j.1523-1739.2003.01491.x

Spencer, F. and Montero Salvatico, Y. (2019) *The guide to the Natural Foresight® Framework: a comprehensive approach to applying strategic foresight in organizational and personal contexts* [Online] Available at: https://tfsx.com/guide/. Accessed 13/06/2025.

Sutherland, W. J. and Woodroof, H. J. N. (2009) The need for environmental horizon scanning. *Trends in ecology & evolution* 24:10, pp. 523–527. https://doi.org/10.1016/j.tree.2009.04.008

Voros, J. 2017 *The futures cone, use and history voroscope* [Online] Available at https://thevoroscope.com/2017/02/24/the-futures-cone-use-and-history/. Accessed 13/06/2025.

Wack, P. (1985) Scenarios: Uncharted waters ahead in *Harvard Business Review* 63:5, pp. 85516, pp. 73–89.

Yüksel, I. (2012) Developing a multi-criteria decision making model for PESTEL analysis. *International Journal of Business and Management* 7:24 pp. 52–66 https://doi.org/10.5539/ijbm.v7n24p52

9 Reimagining stakeholder mapping

By failing to prepare, you are preparing to fail.
(Attributed to Benjamin Franklin)

Key steps

- Change mindset.
- Move on from the matrix.
- Think about the social value of the project and stakeholders.
- Ask stakeholders for their perspectives.
- Think creatively about how to engage.
- Embrace the mess.

Stakeholder mapping is the answer to one of the biggest challenges in project management: how do I work out how to engage with everyone when engaging with everyone is just too much?

This chapter will explain how you can determine whom you need to engage with, and we are not just talking about the usual suspects here. We will also explain how you can shake up the very traditional process of stakeholder mapping by applying a social lens, thus making it more people focussed.

Let's look at how we can carry out stakeholder mapping for a socially successful project.

9.1 Why stakeholder mapping?

The first step in stakeholder mapping is to think about who is important to your project (Bourne and Weaver, 2010). This usually results in a long list of stakeholders. In our experience, most project teams start down this track and end up with a short list of the 'usual suspects', which is too narrow and very inward looking, missing some key perspectives from those who are most impacted. Or a long, baggy list of people, organisations, and categories usually sitting in an unwieldy spreadsheet.

DOI: 10.1201/9781003659877-13

Often this spreadsheet is left on the virtual shelf because it's difficult to know what to do next. It is often overwhelming, especially for non-engagement professionals to even begin to think about how to engage with all the people on the list within the constraints of the resources, budget, and time available on the project.

How do you engage a list of a hundred stakeholders with a team of two people, a small budget, and next to no time? It's a scary place to be.

The only real answer is to segment the stakeholders and make the list more manageable. This is usually done by some form of stakeholder mapping (Kujala et al., 2022) which enables you to move from that long list to a more manageable set of categories prioritised to help you decide where to focus your attention and resources (Boesso and Kumer, 2009).

9.1.1 How it's done

Project managers and consultants love to reach for a four-box matrix, and this tool is most often used as a first step in stakeholder mapping (Mather et al., 2007). This particular four-box matrix was first introduced in 1998 and is often referred to as the interest/influence matrix. It puts stakeholders into the relevant boxes based on two factors – their *interest* in the project and their *influence* on the project (Johnson, Scholes, and Whittington, 2008). Based on each box, you identify how often and by what means to engage with those stakeholders and subsequently create a nice and simple stakeholder engagement plan (Mather et al., 2007). The UK Government Communication Service (2021) promotes this approach in its guidance on effective stakeholder engagement, one of many examples where the two axes creating the four boxes are interest and influence (or sometimes interest and power).

This leads to dividing stakeholders into four groups (Figure 9.1):

You then plan how to engage these four different groups separately. The descriptors from the UK Government Communication Service (2021) suggest:

> Low interest/low influence: **Keep informed** – these are the stakeholders you will supposedly need to spend the least amount of time on, although you should still keep them informed. Once you have plotted your stakeholders

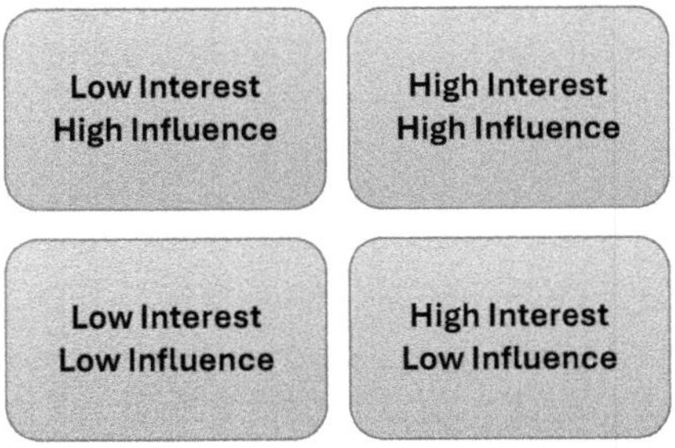

Figure 9.1 The influence/interest matrix.

Source: UK Government Communication Service (2021).

you will be able to prioritise them and identify where and how you should concentrate your efforts.

High interest/low influence: **Consult with** – keep these stakeholders adequately informed and listen to and engage with – often these stakeholders are happy to take on finer details and can be a key sounding board. If you can involve them then they will supposedly act as ambassadors.

Low interest/high influence: **Keep satisfied** – make sure there is awareness, understanding and support for your project among these stakeholders. Put in place enough engagement tactics to ensure that they are informed and satisfied, and that they are kept updated, but do not push or bore them with your messages.

High interest/high influence: **Engage with closely and influence actively** – these are the people who you must fully engage with. They can supposedly have the biggest impact on your project. They may prove to be trusted advisors, key to developing policy and should be seen as partners whose opinion is considered valuable in regard to the project.

Set out in black and white here, how does this approach make you feel? For us, it feels very simple, very transactional, very corporate. It hardly suggests a focus on the people who are going to be impacted by the project, and this is why we have a problem with it.

9.2 Beyond simplicity – adding the social value lens

This approach to stakeholder mapping provides a deceptively useful and certainly simple framework. It's very helpful to move from the long list of potentially dozens of stakeholders to a set of categories. Above all, it gives you the justification to focus more on certain stakeholders, saving time and energy and allowing you to prioritise what's really important about your project … such as the technical, economic, or environmental factors! (slight sarcasm here).

Our dissatisfaction with this interest/influence matrix was a primary reason why we started to advocate for the social dimension of projects.

The fundamental flaw in this framework is the concept of influence or power. By definition, the decision about how much power/influence a stakeholder has sits with the individual or group doing the analysis – which is usually the project team. Stakeholders are given power, or they aren't given power by project teams. Typically, this means that project teams identify familiar stakeholders as having influence, because they are familiar or have always had influence.

9.2.1 Who has power and influence?

The terminology of the interest/influence matrix is extremely power laden. It does not incite an ethos of collaboration. It is very one directional and inward looking. We discussed Arnstein's ladder (1969) in Chapter 4 and the language found at the bottom of that ladder of engagement, depicting tokenistic engagement, is

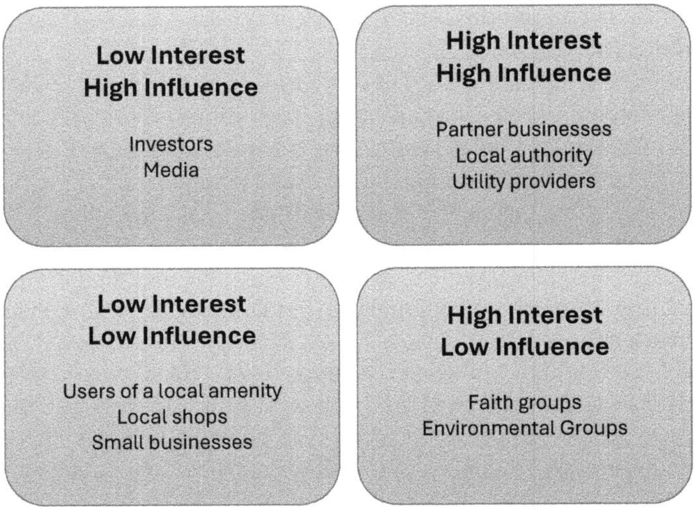

Figure 9.2 Interest/influence matrix with worked examples.

Source: Authors' own work.

very similar to how the matrix is often used. It does not offer hope that the people impacted by the project will be considered sufficiently, let alone have the power to shape and influence the design.

A frivolous example: a dog lover might decide that dogs will have a strong influence on their project. Stakeholder segmentation fulfils that prophecy as dogs become part of a 'high influence' segment given attention by the project team. This example is bad news for cats and cat lovers, even though they may have some very interesting insights to contribute to the project.

In real life, this is bad news for groups and individuals in society who have little power at present and are often the seldom heard or disengaged groups of stakeholders (Bal et al., 2013). This approach risks putting them at further disadvantage by relegating them to a 'low influence' segment that receives little attention from the project. Your approach to stakeholder mapping has a major implication for equality, diversity, and inclusion.

Frequently, the interest/influence segmentation results in the high-interest/high-influence segments being populated by organisations that already know a lot about an issue (hence the high interest) and who are already known and respected by the project team (hence the high influence). Often these organisations (seen as partners whose opinion is considered valuable) are similar to the organisation running the project: they speak the same language and share the same values. For projects run by public sector organisations, small businesses, community groups, and vulnerable groups in society are unlikely to find a place in the favoured top right box of Figure 9.2. Instead, they often find themselves in either of the bottom boxes, with little opportunity to shape a project that may well impact them the most.

Essentially, the current stakeholder mapping tools do not help projects to consider the social dimension. This is a problem, because stakeholder mapping should be one of the first activities in a project (Mather et al., 2007). If it excludes those most impacted by the project, how can you expect the rest of the project to have a social focus?

9.2.2 Using a social lens in stakeholder mapping

Using a social value lens offers new perspectives and helps you identify the social values of your stakeholders.

Let's imagine how the segmentation for a project might play out if the two axes were 'emotion' and 'exclusion', so the matrix is people focused. In Figure 9.3 the top right box would be for stakeholders with high emotion and high exclusion from the project.

Typically, the top right box gains the most attention. In the traditional model, high-interest/high-influence players become key partners who receive the majority of time and attention during engagement and their opinions are considered the most. The bottom left box (low interest/low influence) is often largely ignored or directed to a website or list of frequently asked questions. Traditionally this means that the people most affected by a project are given little power to influence project outcomes. Using this flipped approach, these people are given the power to shape and influence the project – cool, hey?

This is an imaginary example. It is important to remember that each project is unique with its own unique set of stakeholders and therefore the stakeholder mapping outcomes will always look different for each project.

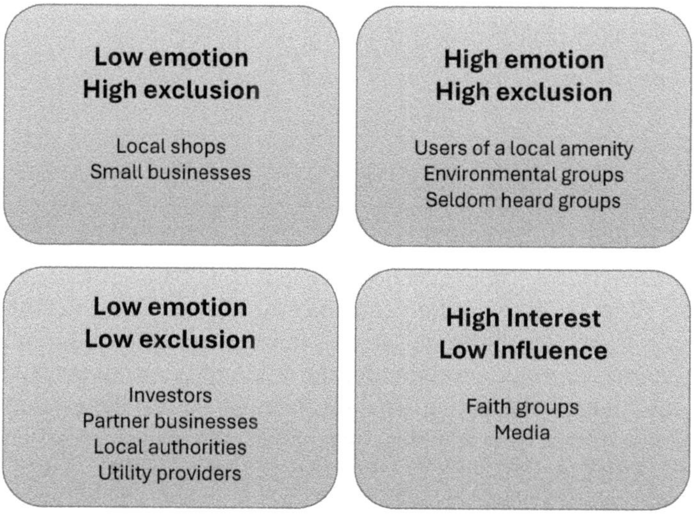

Figure 9.3 Alternative stakeholder segmentation.

Source: Authors' own work.

9.3 Implications for how you engage

The second example – the matrix reloaded in Figure 9.3 – would lead to a completely different engagement profile to that for Figure 9.2. It would involve outreach beyond the 'usual suspects' and require the project team to go to places they don't usually go and engage in a way they don't usually engage, in order to treat users and seldom heard groups as trusted partners. Rather than traditional business meetings with like-minded professionals (who will probably agree with what is being proposed), this would require more time spent in community venues, engaging empathetically with people who might be usually disempowered.

Stakeholder mapping in this way may reveal a need to focus on communities, small community organisations, and small businesses. Engaging them will be more of a challenge, but the value of that engagement will be higher, yielding deeper and more challenging, but valuable insights. Engaging with these people helps you understand what they most value and what they want to see from the project, enabling you to shape a socially successful project.

The other side of this coin is that it's harder to engage or collaborate with people and organisations who are not like us, who may live, think, and work differently from us. You may need to challenge your current practice. For instance, people and groups who are traditionally excluded may require us to go to where they are rather than coming to meet you where you usually go. Indeed, some of these groups are sometimes called 'hard to reach' or 'seldom heard' when really, it's just that we are not good at reaching them. They are not the problem, we are (McEwan et al., 2022). We don't adapt our practices to meet their needs, we expect them to adapt to meet our needs and when they don't, we blame them for not wanting to be involved in the project.

The following case study describes how the Sister project in Manchester, UK, used an innovative approach to stakeholder mapping to shape its engagement approach.

Case study: Sister Manchester, UK

Sister is an innovation district in Manchester City Centre. It is renovating the former University of Manchester north campus and UMIST sites. Sister aims to bring all levels of innovation organisations together to create a thriving ecosystem from large multinationals to small start-ups (Sister Inc, 2025a).

As part of the Sister development, the first building to open in Autumn 2024 was the Renold Building. This was brought back to life as an innovation hub for science and technology innovation entrepreneurs, start-ups, and spinouts (Sister Inc, 2025b).

Useful Projects (Useful Projects, 2025) and Aurora Engagements (Aurora Engagements, 2021) were successful in the tender process to develop the Social Value Strategy for the Sister development in early 2024.

Useful Projects and Aurora Engagements ('the team') presented their philosophy for this project at the tender interview. They wanted the Social Value Strategy to be developed by the people who would use the Renold Building and those living in the surrounding area. The client, Bruntwood SciTech, fully supported this approach to the project.

Stakeholder identification and mapping were important aspects of the project. The team worked closely with a stakeholder engagement consultant to understand what engagement had already been done, what data was available, and who had already been engaged. They then sought to identify more stakeholders and grow the list. Once the list had been compiled, the stakeholders were mapped using a similar tool to the interest/influence matrix but using social criteria. The matrix was reimagined a number of times to understand how the stakeholder engagement needs could change.

Using the data from the stakeholder identification and mapping exercises, a community panel (sometimes termed citizen panel) and a youth panel were convened to help create the Social Value Strategy. The first panel members were identified from the mapping exercise, with others gathered by snowball sampling or asking current members if they knew people who would be interested in joining.

The panels brought together local businesses, local groups, residents, and people who worked in the area. The purpose of the panels was to:

- Bring together representatives of community groups and local communities to update and play a role in plans for the initial phases of Sister.
- Ensure that the team heard a range of views and perspectives, including those less likely to be heard during public consultations.
- Provide a forum for local community representatives to share updates and identify opportunities for collaboration.
- Help shape the Social Value Strategy to ensure that Sister addresses local needs to make a positive contribution to the community in the long term.

The vision was that the panels would continue after the Social Value Strategy had been completed and act as trusted friends during future phases of the Sister project.

The panels were used as a test bed for the ideas and approaches in the Social Value Strategy. Findings from the local needs assessment were presented to the panels to understand how accurate these ideas were and to identify what was missing. This ensured that the Social Value Strategy was concentrating, not on what the team thought was important, but on what the panels saw as important for them and the groups they represented.

In a second phase, the team proposed actions to implement the strategy. The panels didn't always agree with what the team presented. There were many occasions when the team went away from panel meetings to rework what had been presented and to do further work. An example of this was how the team had termed the three key themes for the strategy. The first theme was all about being part of the Sister community and the wider community. The team initially presented this as 'Stitched In', representing the ingrained nature of the people who would use the Renold Building. However, the panels felt this language was very corporate and not accessible to everyone. Through discussions with the panels, this theme was renamed 'Connected Communities'. The time required for this was built into the programme and budget.

The result was a Social Value Strategy that not only met the client's brief, but was influenced, shaped, and co-produced by those who would be using and impacted by the Renold Building (Figure 9.4).

Figure 9.4 The Renold building as part of the Sister development.

Source: Sister (2025).

Lessons from stakeholder engagement perspective

- Engaging early in the project, with support from the client, meant a good relationship was developed with panel members which would serve not just this phase of the project, but future phases.
- Taking time to identify and map stakeholders, with a social value lens, meant the right stakeholders were engaged and helped to create an authentic and effective strategy.
- Don't assume you know what stakeholders want or know.

Lessons from a social value perspective

- Without early engagement and collaboration with the panels, the Social Value Strategy could not have been created as effectively as it was.
- The local knowledge brought by the panel members was invaluable in shaping a Social Value Strategy that reflected the true needs and wants of those using and impacted by the Renold Building.
- Professionals don't always know what is important in terms of social value for communities. The team found that some of their initial ideas on what was important to local communities were incorrect.

Takeaway

Taking a social value lens from the beginning of the project – indeed even before contract award – greatly strengthened the Social Value Strategy.

Figure 9.5 An example office space available in the Renold building.

Source: Sister (2025).

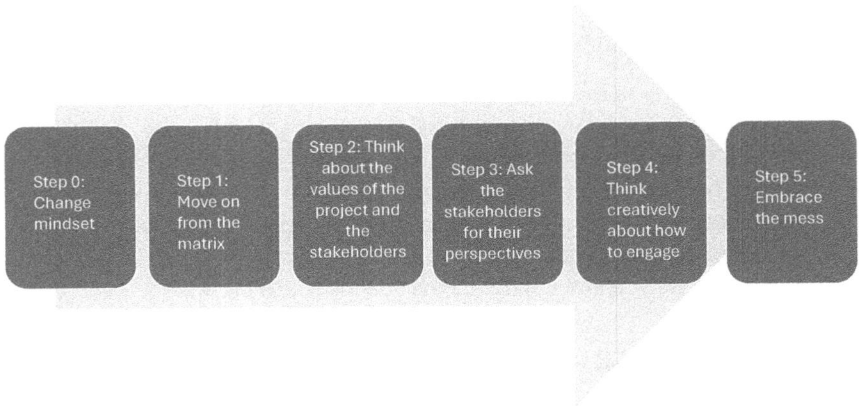

Figure 9.6 Step-by-step guide to the stakeholder mapping process.

Source: Authors' own work.

You may need to find trusted intermediaries or people who are trusted by these groups and involve others in getting your message across and building relationships. This takes time and resources which again need to be planned for.

Above all you need to be ready to hear views that may be unfamiliar and uncomfortable. These are what is going to make your project socially successful. Use these insights, local knowledge, and the social values people have expressed to shape and influence the design of the project.

9.4 Doing socially successful stakeholder mapping

So, what approaches to stakeholder mapping help guarantee socially successful projects? In the last part of this chapter, we want to set out a step-by-step guide to help you shake up your stakeholder mapping process (see Figure 9.6 for an overview). What works for one project won't for another. However, these principles can provide a strong foundation for a socially successful project.

9.4.1 Step 1: It's time to move on from the matrix

Even loading different factors on the x and y axes of the stakeholder mapping matrix still produces a crude categorisation which can miss important nuances and groups of stakeholders.

Matrices can help manage who to engage with and how but should be seen as part of an evolving and living resource that is flexible and continually monitored. It could be that a matrix is used on a number of different occasions and at different stages of the project, with a different focus each time. There is no rule that the same matrix has to be used for the duration of the project. You might change the factors chosen as the project evolves. Different groups of stakeholders will be influential at different stages of the project, and on different matters, so reflect this in your stakeholder mapping.

9.4.2 Step 2: Think about social value both ways

What social value does the project want to create? What social value do stakeholders see the project creating? What do they want and need? Do both perspectives align?

Here it may be useful to think about how different types of knowledge can add value to your project.

> Communities and stakeholders hold other kinds of knowledge about the way things work. These can include local trends and changes, how the environment is used and who uses it, key locations for particular activities, cultural meanings, and feelings about the place.
>
> (Pound et al., 2016)

Different types of knowledge might include:

- Technical knowledge from other disciplines – for example, understanding how emergency planning procedures might affect how you design access routes.
- Organisational knowledge from across organisations – for example, how your project might affect and be affected by local authority policies not just for planning but also for flood resilience or digital inclusion.
- Knowledge of the community – for example, what trusted intermediaries can tell you about community history and hopes, and how they might react to your project or contribute to improving it.
- Environmental knowledge in the broadest sense – this is beyond understanding how to mitigate environmental risks to thinking about natural capital or biodiversity impacts and enhancements.
- Cultural knowledge – what people can tell you about the range of issues that are important to communities and how they may complement each other in the project and even have the potential for conflict.

9.4.3 Step 3: Ask the stakeholders

From the stakeholder mapping exercise, identify a few stakeholders who you have an existing relationship with or who will be willing to engage early. Bring together this group and ask them.

- Whose voices also need to be heard by the project?
- Who isn't on the stakeholder list?
- What different types of knowledge are needed?

Keep asking these questions so your engagement is like throwing a pebble in a pond with the waves reaching deeper into the community. This should not be a one-off. Building relationships with stakeholders over time is a key building block of socially successful projects.

9.4.4 *Step 4: Think creatively about resourcing*

When we share this thinking with project managers, they often raise their eyebrows at the thought of the time and energy this takes.

Often engagement activities are given a small time frame and a small budget and may be the first thing to be trimmed when project cuts are implemented (Bice et al., 2019). However, you don't have to do it all yourself.

Ensuring a stakeholder engagement professional is part of the project team from project inception will help massively. They will help plan, identify risks, and advise what methods can be used. But also use those trusted stakeholder contacts. Ask them to speak with other stakeholders and spread the message to raise awareness of your project. This can be daunting as the power shifts and you cannot control the detail of the messaging, but often new conversations arise which you were not expecting and can be immensely valuable. Which leads us to…

9.4.5 *Step 5: Embrace the messiness*

The understandable impulse to create a neat stakeholder map with a clear list of actions works… until it doesn't. A tight focus and message discipline run the risk of groups and individuals feeling excluded (because they are) and concluding that the only way to have an influence is to kick up a fuss (because it is). So, a stakeholder engagement map and plan may look more like a mind map or spider's web, than a neat Gantt chart or table.

It will need to be adaptive, flexible, and iterative.

As we have said already (probably on numerous occasions throughout this book!), stakeholder engagement isn't a tick box exercise, it's not neat and it's not a linear process. It should be ongoing and proactive. It can't be 'done' then neatly put away in the project file. This is why the stakeholder mapping process we are proposing requires a mindset shift by the whole project team from the beginning of the project.

9.4.6 *Step 0: Mindset shift for socially successful stakeholder mapping*

So how do you, as a project manager, engineer, planner, architect, or engagement officer, manage the mess? It means letting go of the desire for neatness and control and often persuading the senior people responsible for the project, such as your project board, that messiness doesn't mean that things are out of control.

Keeping the engagement objectives clearly in sight and being clear about how you will measure progress against them helps.

Stakeholder mapping is a useful tool to make sense of a complex picture, but sense-making will continue for the duration of the project. Remember that your stakeholders are human beings, with their own social values. And remember that they can help you if you ask them.

Key takeaways

Throw out the traditional stakeholder mapping process if you've been using it.

Turn the matrix on its head and create stakeholder segments that are outward looking, focussing on those who are going to use the project and will be impacted by the project, not those designing and investing in the project.

Remember that even then the matrix is only a starting point.

Stakeholder mapping is a flexible, responsive, evolving activity – it can't, and shouldn't, be completed and left.

Embrace the messiness and be open to the valuable insights a range of stakeholders can provide you and the project, bringing their social values to the centre of your planning.

References

Arnstein, S. R. (1969) A ladder of citizen participation. *Journal of the American Institute of Planners* 35:4, pp. 216–224. https://doi.org/10.1080/01944366908977225

Aurora Engagements (2021) *Aurora engagements* [Online] Available at: https://auroraengagements.co.uk/. Accessed 03/07/2025.

Bal, M., Bryde, D., Fearon, D. and Ochieng, E. (2013) Stakeholder engagement: Achieving sustainability in the construction sector. *Sustainability* 6, pp. 695–710. https://doi.org/10.3390/su5020695

Bice, S., Neely, K. and Einfeld, C. (2019) Next generation engagement: Setting a research agenda for community engagement in Australia's infrastructure sector. *Australian Journal of Public Administration* 78:2, pp. 290–310. https://doi.org/10.1111/1467-8500.12381

Boesso, G. and Kumar, K. (2009) Stakeholder prioritization and reporting: Evidence from Italy and the US. *Accounting Forum* 33: 2, pp. 165–172. https://doi.org/10.1016/j.accfor.2008.07.010

Bourne, L. and Weaver, P (2010) Mapping stakeholders. In: E. Chinyio & Olomolaiye, P. (eds.) *Construction Stakeholder Management*. Blackwell Publishing: Chichester, pp. 99–120. https://doi.org/10.1002/9781444315349

Government Communication Service (Updated 2021) *Ensuring effective stakeholder engagement* [Online] Available at: https://gcs.civilservice.gov.uk/publications/ensuring-effective-stakeholder-engagement/. Accessed 25/05/2025.

Johnson, G., Scholes, K. and Whittington, R. (2008) *Exploring Corporate Strategy* (8th Edition). Prentice Hall: Harlow.

Kujala, J., Sachs, S., Leinonen, H., Heikkinen, A. and Laude, D. (2022) Stakeholder engagement: Past, present, and future. *Business & Society* 61:5, pp. 1136–1196. https://doi.org/10.1177/00076503211066595

Mather, V. N., Price, A., Austin, S. and Moobela, C. (2007) Defining, identifying and mapping stakeholders in the assessment of urban sustainability. *Proceedings: SUE-MoT Conference 2007: International Conference on Whole Life Sustainability and its Assessment*, Glasgow, Scotland, 27th-29th June 2007.

McEwan, L., Holmes, A., Cornish, F., Leichenko, R., Guida, K., Burchell, K., Sharpe, J., Everett, G. and Scott, M. (2022) Rebuffing the "hard to reach" narrative: How to engage diverse groups in participation for resilience. *Journal of Extreme Events* 9:2&3, pp. 1–14. https://doi.org/10.1142/S2345737622500038

Pound, D., Reed, M., Armitage, L., & Pound, J. (2016) *Engaging and Empowering Communities and Stakeholders in Rural Land use and Land Management in Scotland.*Scottish Government: Edinburgh.
Sister Inc (2025a) *Our story* [Online] Available at: https://sistermanchester.com/about-us/our-story. Accessed 03/07/2025.
Sister Inc (2025b) *Renold innovation hub* [Online] Available at: https://sistermanchester.com/the-place/renold-innovation-hub. Accessed 03/07/2025.
Useful Projects (2025) *Consultancy services* [Online] Available at: https://usefulprojects.co.uk/. Accessed 03/07/2025.

10 Optioneering

> Truly successful decision making relies on a balance between deliberate and instinctive thinking.
>
> (Gladwell, 2005, p.141)

Key steps

- Involve stakeholders in developing options.
- Use social criteria in options appraisal.
- Involve stakeholders in options appraisal.
- Do a sense check of your preferred option.

Optioneering takes us to the sharp end of stakeholder engagement: what might we do? And what do stakeholders think of the possibilities?

This chapter looks at how you can develop different options for your project and how to appraise those options, not just through a financial, technical, or environmental lens, but through a social lens.

10.1 The sharp end of engagement

Optioneering takes you from idea generation to the verge of a decision, a final option. It's a classic 'diamond' or diverge/converge activity.

Figure 10.1 draws on the work of Sam Kaner (2014). It shows the progression from the 'diverge' phase, when you are looking to get all the ideas on the table including the essential creative thinking that can open up new options, to the 'converge' phase, when you come together with a set of options, and finally decide on a design or approach.

In the middle of the diamond is what we have genteelly called 'processing' and Sam Kaner (2014) sometimes called the 'groan zone': that phase where you are trying to make sense of the ideas, knowledge, opinions, thoughts, and information

DOI: 10.1201/9781003659877-14

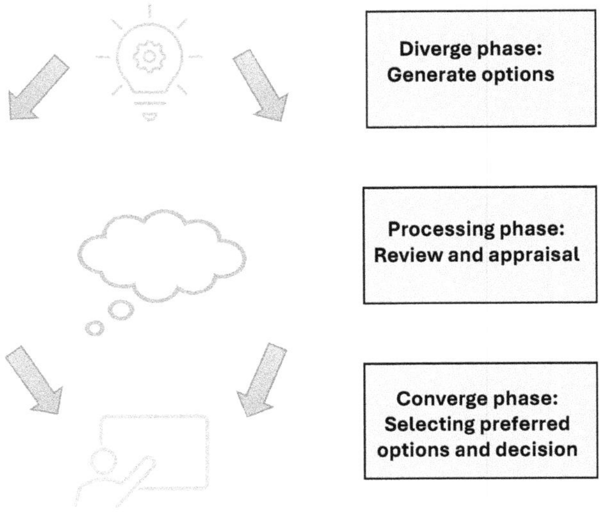

Figure 10.1 The optioneering 'diamond'.

Source: Adapted from Kaner (2014).

gathered. The next step is, bit by bit, to convert it all into possible options, one of which will go forward as your preferred option.

Of course, it's more complex than that. Most projects will have several rounds of optioneering which may cover:

- The big picture fundamental choices: where will we locate that pumping station?
- The mid-level aspects: what technology should we use?
- Smaller but still essential questions: what colour shall we paint it?

10.1.1 Whose options are they?

At the beginning of a project, it is good practice to assign time to generating options and alternatives.

In Chapter 9 we described the messy process of stakeholder engagement. Ideally options emerge from an iterative process of stakeholder engagement including the 'groan zone' where options are explored. More often, stakeholders see less of how the options are created and tend to be presented with a few final options without being told how this position was reached.

The risk is that you miss knowledge and ideas around the actual products for which you are considering options. For instance, community members might have ideas about how a pumping station could become a local attraction connected to a nature reserve, an idea you might have missed and not included in your options.

Involving stakeholders in 'groan zone' discussions will lead to a more robust range of options, more likely to gain approval (Kaner, 2014). In particular it helps them understand both opportunities and constraints, reducing the number of *why don't you just…?* comments further down the track.

George Berry, Architect and CEO of Berry Architecture in Canada echoes this sentiment (Berry, 2025). Berry Architecture does a lot of work with First Nation communities in Canada, and it is important to understand their culture, history, and lifestyle. Creating options for a project without taking these considerations on board would result in a completely irrelevant project that didn't meet the stakeholders' needs. The project team takes the time to engage with the stakeholders and First Nation communities, understand their needs, create options based on these needs and then discuss how the options could be revised based on stakeholder opinions.

You may find that stakeholders are more interested in some aspects of your project than others, and these options will draw the most attention. For the pumping station example, you might expect that community members would be interested in options for location and visual appearance but be less concerned about the technology to be used. We write 'you might expect…'. It's important to ask, *not* assume. Choice of technology may be very interesting to community members because of implications for noise, emissions, or general concern for a low-carbon approach. Stakeholders might bring previous experiences of being promised a technology fix that did or didn't work and provide helpful challenge and support.

Beware of offering choices to stakeholders that are trivial or tokenistic. If you only ask a community about the visual appearance ('what colour should we paint the doors?') they may feel fobbed off. The project team may need to get real at this stage (if not earlier) and consider what influence they are ready to give to stakeholder views.

We worked on a flood management project which involved residents losing part of their gardens, which backed onto a river, to a flood wall. They were offered a choice of different types of railing to top off the wall. This went down badly as they felt that they were being asked about the least important aspect of the project but had no choice over what mattered to them and their families.

As we have already said many times in this book, transparency, honesty, and listening are the founding principles for the social dimension of projects.

10.1.2 Options appraisal

Most options appraisal assesses options against criteria. There are two main types of criteria:

- Screen criteria: if an option does not meet this criterion it will be screened out.
- Sort criteria: options are assessed against this criterion, usually using some sort of scoring process.

The options are then assessed against the criteria to provide an indicative ranking of options. Table 10.1 shows a simplified example: how to choose a holiday destination for a family based in the UK. The numbers signify how well the option aligns with the criteria.

Table 10.1 Assessing options against criteria: holiday destinations

Option/criterion	*Criterion A Total budget no more than £2000. Screen criterion*	*Criterion B Relaxing for adults*	*Criterion C Carbon cost of travel*	*Criterion D Child friendly*	*Total*
Option 1 Paris	Meets criterion	6	5	5	16
Option 2 Seychelles	Does not meet criterion	X	X	X	X
Option 3 Lake District	Meets criterion	9	8	6	23
Option 4 Holiday village	Meets criterion	5	8	9	22

Source: Authors' own work.

In this example, option 2 is screened out because it exceeds the maximum total budget. The other options are assessed against three sort criteria with option 3 receiving the highest score.

10.2 Integrating the social value perspective into options appraisal

It is possible to integrate the social value perspective into this traditional model of options appraisal, but it needs some careful thinking. We believe that this often provides an important practical route to assuring the 'quality' element of a project, alongside 'time' and 'cost' comprising the three 'building blocks' of successful projects (Association for Project Management, 2025b).

10.2.1 Having a social value criterion

The first and obvious point is that you need to have at least one social value criterion as part of the appraisal process. This is by no means a given in many projects. Traditionally, technical, economic, and environmental criteria have dominated options appraisals, which may be described as an 'investment appraisal' using return on investment calculations (Association for Project Management, 2025). In principle this approach should also include 'non-financial benefits' but in our experience these are often seen as secondary considerations. As a first step, we refer back to our overall characterisation of what makes a successful project, so a generic framework of criteria for analysing options might be:

- Is this option technically feasible?
- Is this option financially viable?
- Is this option socially desirable?

This version implies that each criterion is a screen criterion – that is, if you answer 'no' to any of these, the option is discounted.

Alternatively, you could treat these as sort criteria:

- How technically feasible is this option?
- How financially viable is this option?
- How socially desirable is this option?

You could then rate these criteria on a scale, implying that no single criterion has a 'veto'. The downside is that the social is usually seen as least important, with the project team prioritising the technical and financial factors. This could mean going ahead with an option that the team knows has low social desirability.

The criteria need to be real and agreed by the project team as a whole. There is no single model for doing this, but it's important to avoid tokenism in the form of a vague 'review of social impact' at a later stage. Have the team think through how to assess the social value of options to embed that thinking and reassure stakeholders that it is important.

10.2.2 Creating a social value criterion for options appraisal

The examples above are high level and quite simple. More detailed criteria, for instance for assessing options for the visual appearance of a pumping station, are derived from the range of engagement conversations and contributions earlier in the project lifetime.

An example of this was our work on a project to create a mine water treatment plant in an area of natural beauty in the North of England which is described in the following case study.

Case study: River Nent mine water treatment scheme

The River Nent is the most metal-polluted river in the North of England.

The Environment Agency and the Mining Remediation Authority (formerly the Coal Authority) decided to address pollution in the River Nent from historical lead and zinc mines on the North Pennine moors.

A first water treatment scheme has now been built (Nent Haggs) and a second site (Nenthead) was under discussion at the time of writing.

When the schemes were first proposed in 2016, there was strong opposition from members of the local community, residents of an isolated village in an area of beauty and businesses serving tourism.

Opposition from the community took the form of lengthy communication and questions, seeking detailed information about aspects of the schemes such as the impact of the scheme on specific populations such as red squirrels and sand martins. Community members also questioned the need for the schemes and the effectiveness of the solution being proposed.

The technical solution was an innovative 'passive' scheme to treat the water through a series of ponds. This approach had only been used a few times before, and community members questioned whether it would produce bad odours. As the technology was new, it was difficult to give a definitive answer based on past experience. They were also concerned that the ponds would be an eyesore in the bleakly beautiful landscape.

In the early stages, there was limited engagement with the community to seek their views and input. The proposed schemes were largely presented as predetermined designs, which led some community members to feel that the process was happening to them, rather than with them. They valued the local landscape and peace and quiet, which they felt would be damaged by the scheme. They were sceptical about the improvements to water quality which would benefit people living downstream, while all the disruption would fall on them. One community member said that the project team were 'more interested in fish than people'.

Community opposition grew to include lobbying the local parish council and MP, and negative articles appeared in the local paper and on TV.

At that stage, the project team paused the project and called a reset.

Working with stakeholder engagement consultants Wilson Sherriff, they went back to basics with the local community, holding briefings and discussions to share their thinking about the problem they were trying to solve, the significant environmental damage being caused to river wildlife by pollution from the historical metal mines. This included asking the community how they wanted to receive communications. In the first instance, the community wanted information to be sent by post, which was costly and time consuming but ensured that everyone in local postcodes received a letter.

Meetings were held with members of the community, including sessions chaired by the local MP, to discuss concerns and try to find opportunities to address them even if they were not directly related to the mine water treatment scheme, for example, surface water flooding of the main road.

Over time, the community remained largely sceptical about the project, but continuity of team membership meant that there was some grudging respect for what they were trying to do. The project team maintained that the treatment ponds were needed but involved the community in determining where the team would locate them. They shared with the community three criteria for location:

- Technical feasibility.
- Cost.
- Social acceptability.

They involved the community in choosing from a long list of possible sites. The site chosen was different from the original location proposed by the project team, being further from where people lived while still meeting the criteria of technical feasibility and cost.

The adjoining Horse and Wagon field contains the mine water pumping station. Here the project created an informal nature reserve to enhance the environment for wildlife and local residents. The site includes local wildflowers, meadow grasses, and trees along with a wetland to increase biodiversity and remove pollutants from the surface water running off the main road nearby.

Lessons from a stakeholder engagement perspective

- Incomplete engagement with the local community early on delayed the project and meant that more intensive engagement was required later.
- It took a long time to rebuild trust with the community and that was not fully achieved even nearly ten years later.
- Involving the community in the choice of site options was important in ensuring that residents felt that they had a say.

Lessons from a social value perspective

- Community members valued the landscape, quiet, and property values more than environmental benefits.
- Continuity of project team members meant that it was possible to build relationships with community members over time.
- It is challenging to work with stakeholders who experience the downsides of built environment schemes when the benefits are felt elsewhere.

Takeaway

- Giving people a say in the options for site location helped persuade the community that their views mattered.

10.2.3 Social cost-benefit analysis

In the UK, guidance on how to appraise policies, programmes, and projects, which goes well beyond our focus on the built environment, is set out in the *Green Book* (HM Government, 2024). It is based upon ideas of welfare economics and while many of the factors it references are relevant to our discussion, its focus is both wider and narrower than ours.

In summary, the *Green Book* (HM Government, 2024) suggests using a multiple criteria decision analysis lens in the longlisting stage of a project. In the shortlisting stage it suggests a social cost-benefit analysis (CBA) which ‘assesses the impact of different options on social welfare’. This analysis compares the social benefits of different options, including a broad definition of ‘social value benefits’ for public sector organisations and for wider society.

Another example, from the Australian Capital Territory (ACT) Government guidelines for infrastructure projects, outlines possible evaluation criteria for their Wellbeing Framework as follows: access and connectivity, economy, education and lifelong learning, environment and climate, governance and institutions, health, identity and belonging, living standards, safety, social connection and time (ACT Government, 2025).

These guidelines are valuable but tend towards a broad definition of social benefit which may be different from the social value perceived by stakeholders. These also need to be taken into account.

10.2.4 How to assess options against a social value criterion?

Having included one or more social value criteria for option assessment, it can be tricky to work out how to assess against them. Technical feasibility may be complex but it's generally possible to assess based on factors such as ground quality and ease of excavation, distance from sources of power and water, availability of specialist equipment and plant, etc. Most project managers are used to assessing financial viability using a range of tools such as Net Present Value calculations, CBA, and financial risk assessment. These approaches are based on quantifying inputs. Economic, environmental, and technical factors all lend themselves to being turned into quantitative factors. This doesn't align very well with the intangible, qualitative nature of social value as we discuss more in Chapter 12.

We worked on a project considering economic policy options for a government department where options appraisals mainly involved plugging a set of numbers into a spreadsheet. However, even for that project, complexity arose when we began asking stakeholders what mattered most to them, and their opinions often ran counter to what the numbers were indicating.

So, the big question is how do you assess social desirability when there may be a wide and varied range of views among individuals, organisations, and communities potentially affected by your project?

10.3 Involving stakeholders in options appraisal

The answer has to be to involve stakeholders in the options appraisal, ideally having taken them on the journey from idea generation and through the 'groan zone' to confirming the list of options.

At the decision point things tend to 'get real' and we've found that stakeholders who were comfortable in principle with a project begin to push back when it becomes clear that a pumping station could actually be built where the river runs through their park.

Proximity and direct effects are key at this stage. Most stakeholders will be most concerned about what affects them directly so if there's an option that positions the pumping station where they cannot see or hear it, they will probably favour that option. Others may take a broader view with environmental or community groups considering wider and longer term effects of an option.

At this stage stakeholder engagement can be like threading a needle, balancing different opinions so that no one has a veto, but the most broadly acceptable solution can be found. Here the relationships that you have built up during the life of the project become important (see next chapter) with the ability to look people in the eye and have an honest conversation about what matters to them. Expert independent facilitation can support those conversations too.

There is a risk, at this point, of individual stakeholders feeling that their views are getting lost and not heard. You need to consider where social value fits within your appraisal framework and promote this to show that their views are being considered.

10.3.1 Weighting social value

Weighting criteria sharpens the appraisal process by recognising that not all criteria are equal. Here, again, transparency is important, taking people through the process of determining which criteria will be assigned the highest weight when you decide on options.

In our broad framework of technical, financial, and social factors, social value considerations can easily be 'outvoted' by technical and financial considerations. Sometimes this has to happen: the community's preferred location for the pumping station is simply too expensive or technically inaccessible. Where it's that clear cut, the location should not even be considered at the options stage. More likely it's a question of the community's preferred location being more expensive or more technically tricky than another option preferred by the professionals.

Another factor is the number of criteria used. If you have three technical criteria, three financial criteria and only one social criterion you'll bias the appraisal without a high weighting for the social criterion.

It's helpful to think about criteria and their weighting in terms of risk, considering the relative impacts of risks such as:

- Working to the limits of budget with the risk of overruns due to economic shocks or unexpected technical challenges.
- Using less familiar technologies with the risk of setbacks and delays.
- Lack of acceptance by the community with the risk of local campaigns to stop the project and consequent impact on budget, programme, and reputation.

No one said this would be easy, but including social value criteria at this stage is key to ensuring that these perspectives do not get lost as the project approaches the sharp end of decision-making.

10.3.2 Consensus sense check

Using a structured method for options appraisal is useful but it's not enough to feed in the numbers and accept the result that comes out.

It's not quite 'rubbish in, rubbish out' but most options appraisals feature some questionable assumptions including the assumptions and biases of the team making the judgement. They are also easy to manipulate to produce the result favoured by the project team or a vested interest.

A vital final step is to sense check the emerging preferred option. Part of the sense check should be based on risk. Ask: *What could go wrong if we choose this option?*

It's also worth going back to the objectives and goals you set for the project, often many months ago, and consider: *How well does this solution help us achieve our objectives and goals*?

Ideally involving stakeholders will help develop this consensus process and they themselves can act as a sense check for the project.

Consensus aims not for a binary yes/no, but brings in other possible decisions (Kaner, 2014). We feel that usually a third category of decision is enough: *I can live with this*. The aim should be for your preferred option to be supported by your stakeholder group or team with everyone saying either *I support this* or *I can live with this*.

Where stakeholders cannot live with the proposal, ask: *What would it take for you to be able to live with this*? This often points the way to incremental improvements to a proposal creating a final option that everyone can live with.

10.4 How will the decision land?

You will not have universal approval for the option you choose. 'The decision will favour one of the options and, as a result, cut off other options and opportunities for the future' (Association for Project Management, 2025a, p.238). Your decision will land better if:

- You use a social value lens alongside technical and financial considerations.
- You can point to a transparent process for agreeing the criteria for decision-making.
- You involve stakeholders in generating options and making decisions.
- You build and maintain good relationships with stakeholders so you can weather even tough discussions.

Building and maintaining those relationships over a project's lifetime is the subject of the next chapter.

Key takeaways

Contour the stakeholders' journey in the project and involve them in the development of the options, the appraisal criteria and the assessment itself.

Ensure that social criteria are included on equal footing with technical and financial criteria. Do not assume that you know what matters most to stakeholders and communities.

Seek consensus among stakeholders as a preferred option emerges.

Having built good relationships and dealt honestly with stakeholders makes it more likely that they will help you find a good solution.

References

ACT Government (2025) *Options analysis process* [Online] Available at: https://www.treasury.act.gov.au/capital-framework/prove/detailed-technical-guidance/options-analysis/options-analysis-process?utm_source=chatgpt.com. Accessed 24/8/25.

Association for Project Management (2025a) *Body of knowledge* 8th edition. Association for Project Management: London.

Association for Project Management (2025b) *What is project management?* [Online] Available at: https://www.apm.org.uk/resources/what-is-project-management/. Accessed 07/05/2025.

Berry, G (2025) 'Book Interview'. Interview by Sarah Fitton [Microsoft teams] Unpublished, 6/5/2025.

Gladwell, M. (2005) *Blink: The Power of Thinking without Thinking*, Penguin Books: London.

HM Government (2024) *Guidance: The green book (2022)* [Online] Available at: https://www.gov.uk/government/publications/the-green-book-appraisal-and-evaluation-in-central-government/the-green-book-2020. Accessed 24/06/2025.

Kaner, S. (2014) *Facilitator's Guide to Participatory Decision-Making.* Jossey-Bass: San Francisco.

Section 4

The future is bright, the future is … socially successful projects

11 Engagement through the project lifetime

It does not do to leave a live dragon out of your calculations, if you live near him.
(Tolkien, 1995)

Key steps

Include the social dimension at each stage:

- Business case stage.
- Tender and procurement stage.
- Design stage.
- Construction stage.
- Operational stage and beyond.

So far in this book, we have guided you through stages or activities to help consider the social dimension of your project, create authentic engagement, and help surface your stakeholders' social values and the social value of your project.

Now let's zoom out a little and look at how we take all the learning so far and apply it to the project process. How does it differ during the different stages of a project? Should it differ? Do we need to employ different tactics depending on where we are in the project?

We have emphasised the importance of engaging early, and we stand by this principle. However, it is also important to keep engaging throughout the lifetime of a project. Often this does not happen.

Once a project team has completed its design and obtained permissions for construction, it is tempting to cease engagement activities. The team may no longer see the need to engage with stakeholders and there are other calls on the money (if any is left). When the project is built and operating, and the team originally responsible for the project may no longer be involved, stakeholder engagement is virtually non-existent.

DOI: 10.1201/9781003659877-16

But this is when the project becomes real and physical reality begins to affect the lives of the community and stakeholders. Daily life may be disrupted leading to unhappiness, complaints, and even protests, *or* it can be made significantly better, creating benefits and social value for the people who live with the results of the project.

In this chapter we look at how engagement may change during different stages of the project, and why it is important to engage right throughout the lifetime of the project … and beyond. Let's start by looking at a case study that tried to track the change in attitude towards a project over its lifetime.

Case study: Outstrays to Skeffling managed realignment, UK

Managed realignment is a coastal management strategy that reconfigures or removes existing hard defences, such as sea walls, to allow the sea to naturally reclaim certain areas, creating new intertidal habitats like saltmarshes. This approach 'works with nature' to adapt to sea level rise and coastal erosion. Creating new natural defences requires careful planning and can have major effects on people and communities.

The Outstrays to Skeffling scheme was a large-scale managed realignment project in the Humber estuary, Northeast England, that involved building new flood defences further inland and breaching existing ones to allow sea water into the site. This creates new saltmarsh and mudflat habitats for birds, offsets habitat losses from coastal development and other flood risk management activities, and improves flood protection for the area. The scheme was delivered in collaboration by the Environment Agency and Associated British Ports. Construction began in 2021 and much of the work was completed at the time of writing in 2026, with further work to replace aged pumping stations with two new pumping stations underway.

The work created disruption for the local community as it involved closing and re-siting paths, thus changing public rights of way in the area. It involved moving an embankment further inland to establish a new line of flood defences and then breaching the old embankment so that sea water could enter through the breach to create intertidal habitat. On the other hand, it created new habitats for wildlife in the area.

Extensive engagement was undertaken with the community and stakeholders during the design and construction phases of the scheme. This included regular newsletters and web-based information, drop-ins and meetings, a panel of environmental experts, and school visits.

Initial reactions to the scheme were frequently negative with community members unhappy about the loss of farmland resulting from the scheme as well as changes to the landscape and access routes they valued. In particular they were concerned that a proposed new access route denied them views across the estuary, which was something they valued greatly.

These reactions often seemed more important to residents than protection from tidal flooding and improved variety of species. This may have been due to a breakdown of communication between the project team and the community regarding the additional benefits brought by the scheme.

Towards the end of the project in 2025, research was carried out to find out whether there was a process of 'evolution' in terms of local attitudes, from dislike and resistance to the proposals to acceptance and embracing the change.

The responses were mixed: some people's views had not moved on and they remained critical of the scheme, while others were beginning to see positive changes in the habitat and really valued some of the benefits from the scheme, such as improved access.

Lessons from a stakeholder engagement perspective

- It is important for project teams to give people time to digest information where they are proposing serious change and anticipate negative responses without taking it personally.
- Tracking changing attitudes of stakeholders and communities can help understand how opinions evolve and help evaluate the long-term impacts of built environment projects.

Lessons from a social value perspective

- People often value what is familiar and find it difficult to place a value on potential benefits by comparison.
- Emotional responses to change – views and landscapes – need to be recognised as important and not de-valued because they are not seen as important by technical staff. Place attachment and memory are important considerations when proposing changes to landscapes.

Key takeaway

Keep investigating stakeholder attitudes through the duration of the project and beyond in order to understand the full impact of your project.

11.1 Project stage overview

In traditional projects, stakeholder engagement usually happens once the design has been created and before construction has started. Social value is usually considered during the tender and procurement process and then not until the final design has been completed (see Figure 11.1). Both activities are usually compartmentalised into set times or stages of the project process and once they are done, they are done – that part of the project is ticked off.

But this approach doesn't produce a socially successful project.

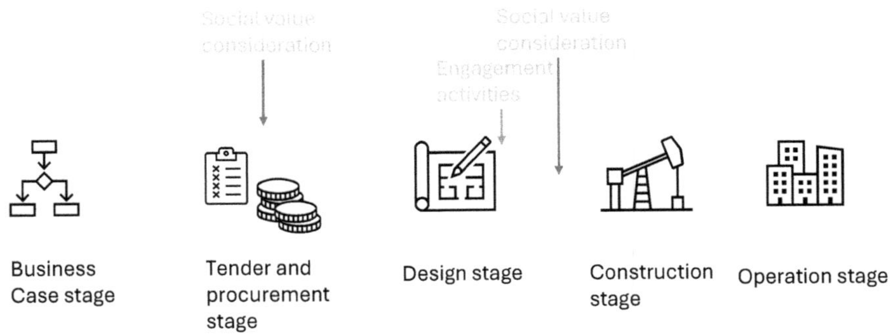

Figure 11.1 Project stages at which social value and engagement are traditionally considered.

Source: Authors' own work.

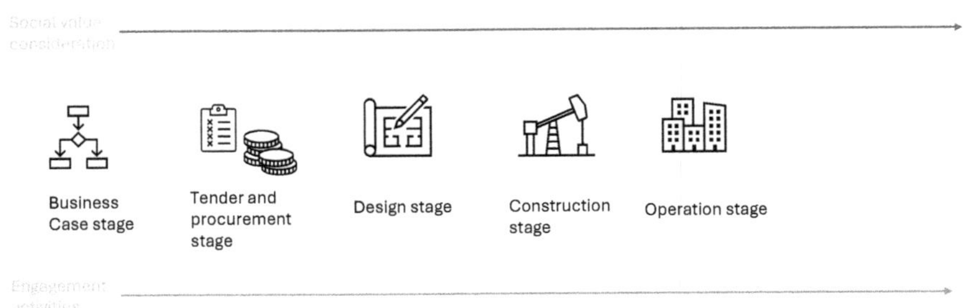

Figure 11.2 Building social value and engagement into the project process.

Source: Authors' own work.

We propose that engagement and social value considerations need to run right through the project from business case development to operation and utilisation, and beyond as Figure 11.2 shows. The UK Green Building Council (2020) explains how the depth of involvement of stakeholders in decision-making during the life of a project correlates to social value opportunities and the likelihood of good social value outcomes. When stakeholders are simply informed of project plans without the ability to influence or shape them, there is a high chance of missing social value opportunities and a low chance of good social value outcomes. However, engaging stakeholders to influence and shape the design leads to fewer missed social value opportunities and higher likelihood of good social value outcomes.

In fact, engagement needs to start even earlier. Figure 11.2 shows the earliest stage: the initial review and business justification and creation of the business case. This is done differently according to the organisation and may be divided into two, including an initial strategic project justification followed by a full business case.

It sounds simple, but why aren't we doing it? And how do we put this new approach into practice?

Let's look at each stage of the project process to understand why this new approach can help us achieve a socially successful project.

11.2 Business case stage

This stage provides the strategic and business justification for the work, and the definition of the problem. The high-level aspirations, or project mandate, are created and discussed at this point to justify any resources going into the tender and procurement stages.

This is the point at which engagement should start, to gain a full understanding of the problem definition and aims and objectives, and to bring the social dimension into play at the outset.

Most projects start with a problem to be solved. For example, how do we prevent tidal flood surges sweeping over the waterfront, making roads impassable and flooding homes and properties?

Don't wait until you have a solution to provide. Share the problem with your stakeholders and include engagement in the business justification stage.

If you don't, you risk trying to sell a solution to a problem that people don't recognise, understand, or buy into. Take them on the journey you have had to go on in order to understand the problem, think about the problem, discuss the problem, then work out the solution. You can't expect stakeholders to jump to understanding the solution when they haven't had the time to digest all the other information (Stewart, 2024). Maybe there have been very few tidal surges till now, but climate change means they will become more frequent: this is the driver for your project. Share this information, explain why the project is needed. Maybe community members think that the occasional flooded road is a price worth paying for peace and quiet. Share the numbers about businesses lost, insurance costs, disruption, longer commuting times.

Sharing the problem in these early stages means listening to what stakeholders have to say. Maybe they will persuade you that the problem is more severe, or less severe, or simply different. Stakeholders may remember previous incidents, perhaps pointing to the knock-on effects on drains and gardens. Businesses may point to the economic damage of a shoreside car park becoming unusable. Stakeholders may remember how the problem was tackled in the past and stimulate fresh thinking. As we discussed in Chapter 8, this local knowledge is invaluable to the project and cannot be gained through desktop studies. The earlier you have this knowledge the more you will be able to use it to shape and enhance the project (Babaei et al., 2023).

Using early engagement to shape and define project objectives offers the best opportunity to get the project definition right from the start.

11.3 Tender and procurement stages

As well as considering social value in the tender and procurement phase, it's helpful to begin thinking about stakeholder engagement at this stage and if you can influence the tender design to include engagement activities in the requirements.

An example was the tender for the design and build of a leisure centre on the UK's South Coast. A senior local authority officer explained how extensive engagement including community engagement, open-door events, and school visits was designed into the tender process and as a result the project delivered things the community said they wanted. The engagement aspect did not have to be shoehorned into the project budget later on.

While tendering and procurement are taking place, engagement can start.

For some of you the idea of engaging with stakeholders before you have anything committed to paper probably fills you with dread. Many of us were trained to engage with stakeholders only when we have all the facts and the answers to every question (Fitton and Moncaster, 2022). There is an element of vulnerability, of saying to stakeholders we don't know about this yet, but this is what we are thinking, what can you tell us?

Demonstrating vulnerability and engaging early helps build relationships, foster trust, and understanding which will take you through the rest of the project (Mulholland, Georgiadou, and Fitton, 2025). Get this stage wrong and you will pay for it further down the line.

And of course this pre-design engagement can feed into the tender and procurement process. Why not involve stakeholders in the choice of contractor? We were a bidder for a project focused on the interests of children and young people. A panel of children and young people interviewed the bidding organisations and advised the customer on their preferred supplier.

11.4 Pre and design stage

Dr Jacqui Cotton (2025), Flood Risk and Coastal Change Research Manager at the UK's Environment Agency, explained how setting clear socially orientated objectives at the beginning of a project helps to not only guide and measure success but ensures everyone – project team and stakeholders – know what the aim of the project is. Too often objectives are not thoroughly debated to expose embedded assumptions as to what success looks like for different stakeholders.

Jacqui (Cotton, 2025) suggested including stakeholders in the governance of the project. Most projects have a very inward-looking governance structure. By including stakeholders in the governance, the perspective of the project is more outward looking and also the project team are held accountable by others rather than just fellow project team members or members of the same or similar organisations. Remember that it takes time to build the psychological safety required for some stakeholders to contribute with confidence.

The stages from concept design to options development and deciding on a design are probably where there is most experience of good practice. But engagement should not end there. Once we have completed a design, stakeholders don't stand still and wait for the final product, the flood waters don't halt, climate change doesn't take a breather. The world keeps spinning, floods keep happening, and stakeholders carry on with their lives, experiencing events, and developing their social values.

Engagement and social value identification and creation should be a continuous process throughout the life of a project (Supply Chain School, 2017).

11.5 Construction phase

The construction stage has potentially the most immediate impact on communities and stakeholders. Residents who have been reconciled to a new development in their village are confronted with a muddy field and traffic noise from 7am for a potentially significant period of time.

Simon experienced this firsthand. A riverside wall had to be replaced, as repeated high flows down the river had broken down the old wall near the main bridge in the centre of the town he lives in. The new flood wall would shore up a series of buildings including shops and pubs, providing lasting protection to 50 homes and businesses. A giant crane was parked on a closed-off section of the main road which allowed materials to be lifted down to the river. The project started off well with good communication from the local and national agencies responsible for the project. A competition to name the crane involved local families who voted for 'Lifty McShifty' (Environment Agency, 2022). Colourful boards were put up around the construction site explaining what was planned and reminding the community of the history of the river and its bridge.

After 18 months of disruption to traffic, having to take a different route to the supermarket and the local station, Simon, and many people he spoke to, was fed up. Even as someone with professional knowledge of the type of project and sympathetic to the problems of construction, challenges caused by extreme weather and unexpected events, he found himself joining in with the grumbling. The information boards were stained and peeling and no one was smiling at Lifty McShifty anymore.

So, what can we do to ensure that we don't lose the trust and relationships we have developed during the design phase, as our projects move into the construction phase?

11.5.1 Realistic expectations

Construction projects always risk running over so the key is to under-promise and over-deliver rather than the other way round.

Before construction starts, explain how the project will affect the community. This may mean focusing on different communities from those most affected by the project itself. It may be that routes for lorries to come into town become clogged affecting the people who live there more than those you've involved most in thinking about the flood gates. It may be that business footfall is impacted by road closures even though the businesses themselves are not near the location of the flood gates. Give people time to prepare for disruption to come.

One of the most difficult aspects of stakeholder engagement is where the benefits of a project are felt in a different place from the disbenefits, so you need to explain who will be affected, why and for how long. If there's going to be noise, dust, or odour, level with people. Give them a realistic sense of the time frame. Emphasise the long-term benefits but don't sugar coat the short-term hassle. Articulate the perceived social value that the project will bring in the future. This is easier if you have engaged early with stakeholders, listened to them and taken account of what matters to them in the project design.

11.5.2 Keep communicating

Project teams often tell us that it's only worth communicating when you have something new to say.

We disagree.

The English footballer Gary Lineker said that to score goals he wasn't 'in the right place at the right time' (Lineker, 2018). He was 'in the right place all the time' so that when the ball arrived, he could put it in the back of the net.

Project teams need to keep engaging even if there is not much new to say, so that the message will get through over time. This involves thinking again about audience, channel, and message (see Chapter 4) given that you may have to communicate with different groups during the construction phase and develop the message over time. This may involve recognising, and apologising for, disruption caused by your project.

People working on the front line may be confronted with questions and comments from the public. High visibility clothing is very… visible. The construction team may suddenly be a key communication channel as well as having their jobs to do while they are on the ground. Investing in briefing or training on how to engage with stakeholders is invaluable for the social success of the project. This is also why we advocate that the social dimension is embedded from the beginning, so everyone knows they have a responsibility to deliver.

Everyone who may come into contact with the public and stakeholders should know what to say and report back to the main project team on what they hear. This reporting mechanism is imperative, so stakeholders feel listened to and heard. You can understand what's worrying the stakeholders, mitigate the issues where possible, and adjust engagement and messaging further.

11.5.3 Find out what matters

As the project evolves, what matters to stakeholders and communities may change (Fitton, 2015). External factors out of your control may influence this as well as the construction activities themselves. Storms during the construction of flood gates to block tidal surges may reinforce the need for the structures and make disruption more acceptable. An economic downturn may make businesses anxious about reduced footfall.

As construction continues, people may be more worried about noise, then odour, then road closures.

Engaging during the construction phase is not just a matter of getting the message out but also listening to what people think and what matters to them (Collinge, 2020). This may be very direct if construction takes place on land belonging to others or to neighbours (such as natural flood management projects).

During Sarah's PhD research (Fitton, 2015) she examined a flood alleviation scheme in northern England. As part of the construction process, two large tubes were brought in to direct the river flow while work was done on an existing weir. Some residents who had lived in the area for many years and knew how the river worked told the construction team the river flow would be too much for the tubes they had brought in. The construction team had done their calculations and were sure the tubes would be adequate for the river flow.

What do you think happened?

During the first intense period of rainfall, the river overflowed the tubes and localised flooding occurred, halting construction and delaying the programme.

Engaging during the construction phase may yield useful information. This should be underpinned by honest and transparent engagement that helps build and maintain trust (Bice, Neely and Einfeld, 2019). Ask community members to work with you on agreeing parameters for construction activities, for instance working hours during which nuisance and noise may be produced, or how to raise issues day to day. Or they may have important local information, as Sarah's example has just shown. Feedback from directly affected community members can help assure your construction process.

11.5.4 Be flexible

Alongside the metrics for progress against your project plans, inputs from stakeholders and the on-the-ground workforce can provide valuable insights that help you refine and improve your project, even at the construction phase. This requires flexibility from the project team. Even during construction plans can, and may need to, change. This goes back to ingraining the social dimension in the project from the beginning and creating the right mindset across all members of the project team.

11.6 Operational phase

Once the project has been built, it's not the end of the affair. During business-as-usual operations stakeholders may still have concerns and still want to have a say. Only now will we know if the project is solving the technical problem it was designed to solve and enhancing the lives of the stakeholders impacted.

This will play out differently for different projects within the built environment, depending in particular on their profile (visibility, odour, noise) in a community and whether, like a bridge or road, they are used by stakeholders regularly.

Sometimes a structure may merge into the environment with little public profile other than memories of how things were before and how it was constructed (like the new flood wall in Simon's town). Other structures are prominent and accepted, disliked, or may divide opinion.

After construction, you need to consider how far stakeholders still need to be involved in discussions about the operations and evolution of the structure as well as if the project is meeting its objectives and 'doing what it says on the tin'.

In the next chapter we will look at monitoring and evaluation of the social success of projects.

11.7 How do social values change during a project's lifetime?

We can all think of the arc of public opinion surrounding major construction projects. It moves from a contested space to a noisy building site, to an eyesore to, finally, a familiar feature of the local scene (with luck). Or it moves from a little understood project to a secret plan, to an eyesore, to a white elephant (with less luck).

Flexible engagement over the life of a project enables you to capture the evolution of social value (Fitton and Moncaster, 2022). Understanding social value increases the chances of success for the current project and future projects you may work on.

Over time, different stakeholders become more vocal and prominent while others fade away, their concerns perhaps allayed or having lost interest. The mood may change as different dimensions of the project come to the fore: cost and waste, appearance and aesthetics, use and exclusion.

People's social values change over time (O'Brien and Wolf, 2010). Experience and context affect our social value – what we like, what we don't like, what we will accept, what we won't accept. In the UK social value around working from home and in offices was transformed by the COVID-19 pandemic. Writing in 2026 we can say that there is a strong social value among office workers around the flexibility offered by working part of the time from home or elsewhere than the office. This may change over the coming years as people's experience, and the external environment, affects that value.

This affects projects too. If the project has a long timeframe, what was accepted at the beginning may no longer be acceptable at the end. You don't want to reach the end of the project and find this out. By continually engaging and being aware of social value, you can monitor changes and adapt the project accordingly (within reason obviously) hence mitigating the risk that the final project won't be socially fit for purpose (O'Connor et al., 2025).

In an ideal world all stages of a project from inception to design, construction, and operations would be informed by stakeholders who become advocates and supporters of what you are seeking to achieve. At a minimum, continuous engagement and awareness of what matters to stakeholders reduces contention which can lead to obstruction and delays (Bice, Neely, and Einfeld, 2019). Usually, different teams are involved in different stages, so continuity and flexibility are achieved by effectively sharing learning through the lifetime of the project (O'Connor et al., 2025). This is why monitoring and evaluation of engagement and social value activities is important for the social success of a project.

Katie Wray, a Director at Deloitte agreed that consistency is important on projects. She explained (Wray, 2025) how projects that she had worked on had been far more successful both from an engagement perspective and a social perspective, if the project team remained the same for the lifetime of the project. This is due to the ability to develop and maintain relationships with stakeholders. This takes a lot of work, time, and resources for a project team and if the team changes, relationships need to be developed again. The impact on the stakeholders is also significant. A change in personnel means they not only have to develop a new relationship but also share their thoughts and opinions again. Project team changes are inevitable and sometimes cannot be avoided, but where possible, and from Katie's experience, consistency is key.

Key takeaways

Ingrain a social ethos in the project from the start.

Champion a change to normal practice and routines and start engagement early, including at project justification stage.

Involve stakeholders in developing objectives and in project governance.

Allow social value to shape and enhance every stage of the project, recognising that what matters to stakeholders changes over the course of a project.

Where possible maintain consistency of personnel in your project team.

References

Babaei, A., Locatelli, G. and Sainati, T. (2023) Local community engagement as a practice: An investigation of local community engagement issues and their impact on transport megaprojects' social value. *International Journal of Managing Projects in Business* 16:3, pp. 448–474. https://doi.org/10.1108/IJMPB-10-2022-0224

Bice, S., Neely, K. and Einfeld, C. (2019) Next generation engagement: Setting a research agenda for community engagement in Australia's infrastructure sector. *Australian Journal of Public Administration* 78:2, pp. 290–310. https://doi.org/10.1111/1467-8500.12381

Collinge, W. (2020) Stakeholder engagement in construction: Exploring corporate social responsibility, ethical behaviors, and practices. *Journal of Construction Engineering and Management* 146:3, pp. 1–14. https://doi.org/10.1061/(ASCE)CO.1943-7862.00017

Cotton, J. (2025) 'Book Interview'. Interview by Sarah Fitton and Simon Wilson [Microsoft teams] Unpublished, 5/02/2025.

Environment Agency (2022) *Huge crane in Matlock named as project reaches one third completion* [online] Available at: https://www.gov.uk/government/news/huge-crane-in-matlock-named-as-project-reaches-one-third-completion. Accessed 22/6/2025.

Fitton, S. L. (2015) *Social Value in Practice: A Case of Flood Alleviation Schemes*. PhD Thesis, University of Cambridge.

Fitton, S. L. and Moncaster, A. (2022) Social value, infrastructure and stakeholder engagement: A complex triangle. *Proceedings of the Institution of Civil Engineers – Engineering Sustainability* 175:4, pp. 194–201. https://doi.org/10.1680/jensu.21.00030

Lineker, G. (2018) *Right place all the time* Available at: https://www.facebook.com/watch/?v=1404892512948031. Accessed 12/06/2025.

Mulholland, C., Georgiadou, M. C. and Fitton, S. L. (2025) Key Challenges in Social Value 2025: responsible stakeholder engagement, robust data management and reliable digitalisation. *White Paper.*

O'Brien, K. L. and WOLF, J. (2010) A values-based approach to vulnerability and adaptation to climate change. *Climate Change* 1, pp. 232–242. https://doi.org/10.1002/eet.1658

O'Connor, R. A., Sanchez, E., Bice, S., Jones, K. and Henderson, H. (2025) Avoiding harm or creating benefit? How a risk focus sidelines social considerations in early decisions for Australian infrastructure projects. *Impact Assessment and Project Appraisal* 43:1, pp. 2–15. https://doi.org/10.1080/14615517.2024.2448415

Stewart, I. S. (2024) Advancing disaster risk communications. *Earth-Science Reviews* 249:104677, pp. 1–14. https://doi.org/10.1016/j.earscirev.2024.104677

Supply Chain Sustainability School (2017) *Social value and design of the built environment* [Online] Available at: https://www.supplychainschool.co.uk/wp-content/uploads/2019/10/Resource-ID-5670.pdf. Accessed 11/06/2025.

Tolkien, J.R.R. (1995) *The Hobbit* (5th Edition), pp. 226. HarperCollins: London.

UK Green Building Council (2020) *Delivering social value: Community engagement hacked* [Online] Available at: https://ukgbc.org/resources/delivering-social-value-community-engagement-hacked/. Accessed 11/06/2025.

Wray, K. (2025). 'Book Interview'. Interview by Sarah Fitton and Simon Wilson [Microsoft teams] Unpublished, 10/03/2025.

12 Monitoring, evaluation, and learning

Since all models are wrong the scientist must be alert to what is importantly wrong. It is inappropriate to be concerned about mice when there are tigers abroad.

(Box, 1976, p. 792)

Key steps

- Consider monitoring, evaluation, and learning early in the project.
- Monitor project progress systematically.
- Evaluate the project against objectives.
- Have a system for continuous project learning.
- Involve stakeholders at every stage.

How do you know your project is a success? With luck you end up with something constructed that wasn't there before.

But beyond that, what has been achieved? Was it achieved on time and within budget? Did it solve the technical problem that prompted the project in the first place? How has it impacted the community? What is the perceived legacy of the project: delight or exasperation?

In Chapter 13, we look at, arguably, the most neglected part of a project: monitoring, evaluation, and learning. We will look at why it's important to monitor and evaluate the success of your project, and not just the easy things that can be counted! We will look at why this is often the hardest part of the project, especially for the social dimension, and provide some guidance on making this a key stage, not just during design and construction, but post completion and during operation.

12.1 What is monitoring, evaluation, and learning?

A project benefits from setting out early the framework for monitoring, evaluation, and learning.

DOI: 10.1201/9781003659877-17

Often, we've seen project teams begin to think about evaluation when the project is well underway. In the rush to get things started, monitoring, evaluation, and learning are seen as issues for later. But as we all know too well, later sometimes never comes… or it comes with an extremely small budget.

These three activities are different but interrelated.

Broadly, ***monitoring*** is the continued collection of data about how the project is progressing. It involves translating project objectives into performance indicators which may have targets attached, usually in the form of KPIs (Key Performance Indicators). Monitoring reports alert project managers to issues to address before the problem impacts the overall success.

Evaluation reflects on what has been achieved. It includes an analysis of why intended results were or were not achieved.

> Evaluation is a systematic assessment of the design, implementation and outcomes of an intervention. It involves understanding how an intervention is being, or has been, implemented and what effects it has, for whom and why. It identifies what can be improved and estimates its overall impacts and cost-effectiveness.
>
> (HM Treasury, 2020, p.5)

Learning builds on both monitoring and evaluation to pay particular attention to how lessons are drawn from what has taken place and embedded in thinking about future practice (Kusek and Rist, 2004).

Once objectives for a project are in place, you can develop a monitoring, learning, and evaluation framework, based on these objectives and the vision of success. This means identifying the measures (KPIs) that will show whether or not a project has succeeded.

These may be quantitative (e.g. how many people attended engagement events) or qualitative in nature (e.g. what did people think about the events).

The UK Government's *Magenta book* (HM Treasury, 2020) provides comprehensive guidance on evaluation rationale and methods. It is particularly useful as a reference for publicly funded projects.

The EFQM (European Foundation for Quality Management) model is another framework which recommends two ways of assessing progress: performance measures and perception measures. Performance measures are the quantitative measurables. Perception measures involve asking stakeholders what they think. This helps check whether your project has met the needs and expectations of stakeholders and uses their views and insights to identify improvements and strategic changes needed (EFQM, 2025). This is where one challenge lies: the tension between qualitative and quantitative data.

12.2 Qualitative data versus quantitative data

One of the biggest challenges for the social dimension of projects is how to measure the social outcomes and success of the stakeholder engagement process. This is down to the qualitative nature of the data (Watts and Higham, 2021). Quantitative data, especially economic or monetary figures, traditionally still dominate all forms of decision-making, including KPIs and monitoring data on built environment projects (Scholtens, 2006).

The data gained from the stakeholder engagement process may include intangible sentiments, feelings, opinions, thoughts, and stories. These cannot be put into figures (Watts and Higham, 2021). The Scottish Land Commission ran a research project in 2020 to understand and evidence the benefits of early stakeholder engagement. Their aim was to allocate figures to the benefits. The research found that it was difficult to assign figures to the benefits of early engagement due to their qualitative nature. Out of all the projects and organisations featured in the research, only one company tried to assign financial figures to the benefits of early engagement (Scottish Land Commission, 2025).

This tension may reflect and amplify tensions between different disciplines or backgrounds within the project team. Sarah, along with Dr Karen Potter from the Open University, researched how good, authentic, early communication in a project can be perceived as social innovation. Almost all the project team members they interviewed saw innovation as big, shiny, technologically oriented outputs. Early, collaborative engagement was not seen as innovative. In fact, because the work of engagement professionals couldn't be quantified, there was some tension as the more technical disciplines questioned the 'value for money' of the engagement team. Their success was judged on the number of people engaged, not the rich conversations they were having. Yet the success of the project hung on the strength of collaboration with the affected communities (Potter and Fitton, 2026).

Similarly, both industry and academia are debating how and why to measure social value (Raiden and King, 2023). Over the years different tools and methods have been tried. The TOMs (Themes, Outcomes, and Measures) framework (Social Value Portal, 2025), Social Value ESG Calculator, the social value engine, EcoMetrics (Social Value International, 2025), and Social Return on Investment (SROI) (APM, 2025) all offer a methodology to measure and quantify social value outputs and outcomes.

Quantitative and monetary measures of value can be included in project decision-making processes (Staniškienė and Stankevičiūtė, 2018; Watson and Whitley, 2017). However, because social values are inherently personal and specific to individual and community experiences and circumstances, they are difficult to express in that way (Arcagni et al., 2021). And the meaning of the numbers is open to question (Gibbon and Dey, 2011). If we provide a figure for the social value of a project what does this mean to the project, client, or community? Numerical measures miss the diversity and complexity of stakeholders and the local contexts (Arcagni et al., 2021).

If we don't quantify stakeholder engagement and social value, we risk omitting it completely from decision-making and monitoring. However, if we do quantify it, we need to recognise the limitations. We are kind of in a sticky situation with this.

Let's look at some approaches for developing a monitoring and evaluation framework that incorporates the social dimension of a project … as well as it can for now!

12.3 Framework guide for monitoring, evaluation, and learning

Following the principle outlined throughout this book, involving stakeholders in monitoring and evaluation can help to assess the social success of a project (Feor, Clarke, and Dougherty, 2023). Figure 12.1 shows a high-level process for monitoring, evaluation, and learning, including how stakeholders can feature at each stage to bring the social dimension to your monitoring and evaluation process.

12.3.1 Involve stakeholders early

Involve stakeholders in defining project objectives and how you will measure and assess whether they have been met. This is an opportunity to embed social value in project objectives. It is also an opportunity to demonstrate transparency, as well as benefit from stakeholders' knowledge. Early and continuous involvement means stakeholders are embedded throughout the project process. What does success look like through their eyes?

> At its core, evaluation should be useful to its many different audiences and users. Building the evaluation design around users' needs will ensure they are engaged and that the outputs are of use to, and used by, them. Stakeholder involvement is strongly recommended as the choice of method affects the questions that can be answered (choice of method may depend on which assumptions are most plausible and stakeholders may have a better perspective than government analysts on this).
>
> (HM Treasury, 2020 p.47)

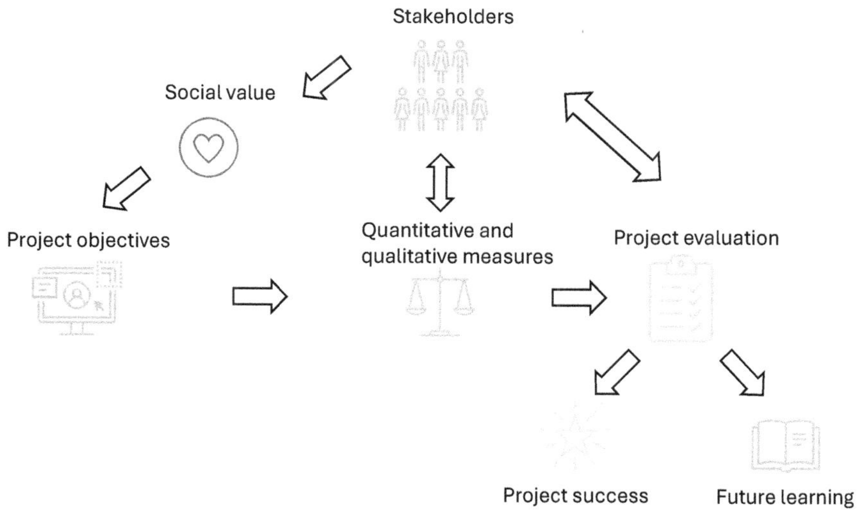

Figure 12.1 Stakeholder-driven monitoring, evaluation, and learning process.

Source: Authors' own work.

For example, a project objective might be 'Achieve a design that chimes with the character, look and feel of the local area'. This reflects social value and is difficult to measure quantitatively. Measures might include 'community members are involved in selecting the design and advising on design elements', 'design is approved by local heritage groups', 'at least 60 per cent of community members support the design when surveyed'.

12.3.2 Involve stakeholders in monitoring

Stakeholder perceptions provide valuable and insightful qualitative data. Qualitative data-gathering tools include interviews, surveys, storytelling, and focus groups structured around the objectives and measures for the project.

It's also worth considering how far to involve stakeholders in the monitoring process itself. Having invested stakeholders either gathering monitoring data or receiving monitoring reports demonstrates transparency and brings additional perspectives. Stakeholders may have ideas about why objectives aren't being met that technically focused project team members aren't able to see.

12.3.3 Involve stakeholders in evaluation

Whose evaluation is it? The lessons of evaluation can be painful, and it may be tempting to keep them private, but transparency affords the opportunity to gain additional perspectives and insights.

Again, we would advocate a dual role for the stakeholders: gaining their insights into the success or otherwise of the project (linking back to the social value discussion); and being part of the evaluation team reviewing progress.

Stakeholders can provide insights into how the project was run, not just what was achieved. Their views are also important in understanding legacy. Was the disruption to the community during construction so demoralising that the new structure is viewed with scepticism; or did the experience of engagement during the project leave a legacy of goodwill? These types of qualitative evaluation enhance the overall framework.

12.3.4 Involve stakeholders in learning

The final step is to involve stakeholders in exploring what you have learnt as a project team. Adding the idea of learning to the monitoring and evaluation process is a fairly recent development. It opens up the scope beyond simply understanding whether a project has succeeded (important though that is) and provides the opportunity to ask:

- What have we learned individually and collectively about the project itself?
- What have we learned that may not relate specifically to the original project objectives?
- What learning can we take into future projects?
- What learning can we share with colleagues within and beyond our organisations?

We worked on a great project involving many different organisations and groups over a period of around 18 months. As the project evolved different groups and organisations became more or less involved. At project close we ran a 'learning sharing event' (which also doubled as an end of project party – bonus!), inviting all the stakeholders to reflect and share learning. The event had a celebratory feel but also unearthed some challenging issues.

Themes from the event were included in the project evaluation report. The learning derived from the project was seen as common property, valued by project stakeholders as well as the core project team. The project team learned about working with communities, and community members learned about how to approach future projects.

Learning is a multi-way street. Jo Dobson, Director at Useful Projects (Dobson, 2025) reinforced this point by calling for more case studies to showcase the good work of socially successful projects. More case studies will help provide an evidence base and justification for this approach but also help others to learn and apply similar thinking on their projects.

12.3 The collective benefits of learning

It's never nice to talk about failure, but this seems like an opportune time to discuss how it is fundamental to learning.

Learning isn't just about looking at what was successful and how to replicate it. Learning why things weren't so successful, or failed, is just as important. We tend not to admit that something hasn't worked as it should. We only highlight 'best practice' as if nothing ever goes wrong on a project… which we all know isn't true!

Understanding what has gone wrong on a project, and why, can help avoid similar situations in the future if the learning is communicated. If learning is not disseminated, the same mistakes are made again and again.

We worked on a project that had engagement at the heart. The ethos was spot on for a socially successful project. We did early engagement, took a social perspective during our stakeholder mapping and instilled collaboration and co-production with the project team. However, when it came to it, we struggled with stakeholder engagement – no one was bothered about the project! We had done everything that 'best practice' had told us, but we had no guidance on what to do when stakeholders don't want to engage. No project admits that the stakeholders don't care. We pulled it around, but we had to learn from scratch rather than the experiences of others.

So, our call to action is this. Shout about all aspects of your project: the 'best practice' but also the bits that gave you sleepless nights or difficult conversations. This learning is just as important. Our case studies should include the good, the bad, *and* the ugly.

A transparent, collaborative approach to monitoring, evaluation, and learning means that learning can be seen as a collective resource, benefiting the wider community of practitioners. In turn it may help prepare the ground for participatory approaches to future projects, extending understanding and commitment among stakeholders.

If a community learns to expect poor communication and having their views disregarded, they are likely to take that learning to future projects. It will reinforce the scepticism many communities feel about being consulted.

On the other hand, a good experience with one project makes it more likely that community members will approach the next one with an open mind, willing to make their contribution and work with the project team to make the best of it.

12.4 Funding ... or the lack of it

So far, we have discussed how we can monitor and evaluate during the course of the project itself. However, we don't know how socially successful a project will be until a few years after its completion once the community has got used to what has been built. At this point, project teams are long gone as is the funding and drive to do any monitoring and evaluation (Kinsbergen et al., 2021).

At that stage there is no way of knowing if a project isn't working for a community (Feor, Clarke, and Dougherty, 2023) – they are essentially stuck with it. Post Occupancy Evaluation (POE) (Watson and Whiley, 2017) tries to address this by gathering feedback and qualitative data on users' experiences anything from six months to years after completion. This allows the project team to assess the social success and what could be changed (if anything can), and more importantly to learn for the next project. POE requires either additional funding or for a project team to be contracted after the completion of construction – something that very rarely happens (Feor, Clarke, and Dougherty, 2023).

This reinforces the need to think about monitoring and evaluation frameworks early in the project process and include them in procurement contracts and budgets.

Academia and research institutes can also play a part. As we have seen throughout this book, social value and many built environment projects span decades and generations. To fully understand social success might require a longitudinal approach and monitoring for years, even decades. Research projects could provide the evidence required to understand, and justify, the need for social success. This is something we don't currently have in either academia or industry. But we need it. We need the data and the evidence to show just how important socially successful projects are.

So, another call to action is for those with the funding and the resources to advocate for long-term studies into the social dimension of projects. One example of this is the partnership between the Quintin Hogg Trust and the University of Westminster, UK. The trust provides funding to the university to help advance the education of students (Quintin Hogg Trust, 2025). We were lucky to work in partnership with the University of Westminster Business School and received a small grant for research into stakeholder engagement in the built environment. As part of the grant, we paid a small number of current students to help with the research which allowed them to forge connections with industry and also develop skills.

All this will take a while, but humanity isn't going anywhere, and future generations will thank us for it.

Key takeaways

Involve stakeholders throughout the monitoring and evaluation process.

Qualitative data is more important than quantitative data when evaluating the social dimension.

Don't underestimate the importance of learning from the data – the good and bad bits. Where possible take the learning from one project to future projects.

Factor into budgets, procurement contracts, and programmes that the monitoring and evaluation of your project doesn't stop when occupation or utilisation starts – this is just the beginning of monitoring social success.

References

APM (2025) *Social Return on Investment (SROI): A powerful tool for the realisation of benefits* [Online] Available at: https://www.apm.org.uk/resources/find-a-resource/social-return-on-investment-sroi/. Accessed 23/06/2025.

Arcagni, A., Fattore, M., Maggino, F. and Vittadini, G. (2021) Some critical reflections on the measurement of social sustainability and well-being in complex societies. *Sustainability* 13:22, pp. 1–16. https://doi.org/10.3390/su132212679

Box, George E. P. (1976) Science and statistics. *Journal of the American Statistical Association*, **71**:356, pp. 791–799, https://doi.org/10.1080/01621459.1976.10480949.

Dobson, J. (2025) 'Book Interview'. Interview by Sarah Fitton and Simon Wilson [Microsoft teams] Unpublished, 27/2/2025.

EFQM (2025) *The EFQM Model* [Online] Available at: https://efqm.org/the-efqm-model/. Accessed 19/06/2025).

Feor, L., Clarke, A. and Dougherty, I. (2023) Social impact measurement: A systematic literature review and future research directions. *World* 4, pp. 816–837. https://doi.org/10.3390/world4040051

Gibbons, J. and Dey, C. (2011) Developments in social impact measurement in the third sector: Scaling up or dumbing down? *Social and Environmental Accountability Journal* 31:1, pp. 63–72. https://doi.org/10.1080/0969160X.2011.556399

HM Treasury (2020) *Magenta book: central government guidance on evaluation.* [Online]. Available at: https://assets.publishing.service.gov.uk/media/5e96cab9d3bf7f412b2264b1/HMT_Magenta_Book.pdf. Accessed 25/08/2025.

Kinsbergen, S., Koch, D-J., Plaiseier, C. and Schulpen, L. (2021) Long-lasting, but not transformative. An ex-post sustainability study of development interventions of private development initiatives. *The European Journal of Development Research* 34, pp. 51–76. https://doi.org/10.1057/s41287-020-00346-0

Kusek, J. Z. and Rist, R. C. (2004) *Ten steps to a results-based monitoring and evaluation system: a handbook for development practitioners*. The World Bank: Washington. https://documents1.worldbank.org/curated/en/638011468766181874/pdf/296720PAPER0100steps.pdf

Potter, K. and Fitton, S. L. (2026) Surfacing hidden innovation: The case of project groundwater. *Journal of Flood Risk Management* 19:1, pp. e70180, https://doi.org/10.1111/jfr3.70180.

Quintin Hogg Trust (2025) *Supporting the advancement of education of students at the University of Westminster* [Online]. Available at: https://www.quintinhoggtrust.co.uk/. Accessed 21/08/2025.

Raiden, A. and King, A. (2023) Added value and numerical measurement of social value: A critical enquiry. *Buildings and Cities* 4:1, pp. 767–782. https://doi.org/10.5334/bc.330

Scholtens, B. (2006) Finance as a driver of corporate social responsibility. *Journal of Business Ethics* 68, pp. 19–33. https://doi.org/10.1007/s10551-006-9037-1

Scottish Land Commission (2025) *The value of early engagement in planning* [Online] Available at: https://www.landcommission.gov.scot/our-work/housing-development/early-engagement-in-planning. Accessed 19/06/2025.

Social Value International *Accredited Software Directory* [Online] Available at: https://www.socialvalueint.org/tools-and-software. Accessed 23/06/2025.

Social Value Portal (2025) *Social value measurement* [Online] Available at: https://www.socialvalueportal.com/measurement. Accessed 23/06/2025.

Staniškienė, E. and Stankevičiūtė, Z. (2018) Social sustainability measurement framework: The case of employee perspective in a CSR-committed organisation. *Journal of Cleaner Production* 188:1, pp 708–719. https://doi.org/10.1016/j.jclepro.2018.03.269

Watson, K. J. and Whitley, T. (2017) Applying Social Return on Investment (SROI) to the built environment. *Building Research & Information* 45:8, pp. 875–891. https://doi.org/10.1080/09613218.2016.1223486

Watts, G. and Higham, P. (2021) If the only tool you have is a hammer, you will treat everything as a nail: Exploring why construction contractors monetise social value. *Journal of Leadership, Accountability and Ethics* 18:3, pp. 10–20. https://doi.org/10.33423/jlae.v18i3.4405

13 Will Artificial Intelligence (AI) sweep it all away?

> It is an important and popular fact that things are not always what they seem. For instance, on the planet Earth, man had always assumed that he was more intelligent than dolphins because he had achieved so much – the wheel, New York, wars and so on – while all the dolphins had ever done was muck about in the water having a good time. But conversely, the dolphins had always believed that they were far more intelligent than man – for precisely the same reasons.
>
> (Adams, 2020 p.110)

AI is everywhere. Most of us use it on a daily basis through our mobile phones. We are seeing our professional lives change as AI changes processes and supposedly makes certain tasks easier or quicker. But what does AI mean for the very humanistic practices of social value and stakeholder engagement? In this chapter we want to briefly explore what AI could mean for the future of social value and stakeholder engagement. We don't have the answers by any means, but we felt we couldn't omit this subject from this book.

We appreciate that some of the comments in this chapter may have been overtaken by events by the time this book is published, but we hope that the approach outlined will still make sense.

As an experiment we asked a well-known AI platform (Large Language Model) to tell us the issues that might arise in terms of stakeholder engagement on a live project. The answer aligned quite well with the approach we were taking. It included a reference to 'cultural and social value'. We also asked about the social value of the community, and again the response matched much of what we had gleaned through human interaction, although it was very generic.

So could AI sweep away, or significantly reduce, the need for human engagement to ensure that projects are socially successful? Could we envisage an AI model that does it all, factoring in the social dimension to produce perfect socially successful projects? We believe that there are opportunities, but we do have reservations.

13.1 Reservations

Using the AI example above, we could use this type of response on a project as a first draft to spur thinking. It might save time. However, the response is very

DOI: 10.1201/9781003659877-18

generic. We would need to dig deeper to reach the seams of understanding required for a real engagement approach.

Gleaning information in this way is not relational and as we have commented throughout the previous chapters, building good relationships is key to socially successful projects.

We can use AI to tell us how much a historic monument is valued by a community. But asking members of the community about what it means to them gives us the opportunity to build relationships with them.

Sometimes the distinction is made between explicit and tacit knowledge:

> "Explicit" or codified knowledge refers to knowledge that is transmittable in formal, systematic language. On the other hand, "tacit" knowledge has a personal quality, which makes it hard to formalize and communicate. Tacit knowledge is deeply rooted in action, commitment, and involvement in a specific context.
>
> (Nonaka, 1994, p.16)

Often, community members have exactly that type of tacit knowledge which has never been written down because it does not lend itself to being written down. It may be linked to the culture, history, and traditions of the place. We understand that the jury is still out on whether AI can provide access to the tacit knowledge that community members have which may never have been expressed formally. For now, we believe that human interaction is key to unearthing tacit knowledge that may be useful to projects.

Above all, you need to know what questions to ask. When we tested our example case, we used our existing knowledge to steer the responses, addressing issues that we already knew existed. So, we would suggest that the person asking the questions of the Large Language Model needs to know about social value and stakeholder engagement as well as having prompt design skills (see below).

On past projects, we have also used AI to build engagement resources. For instance, we used an AI platform to generate visual images to represent 'personas': generic characters who would be affected by a project. We used the visual images as a prompt for discussion about the impacts of the project.

The images produced in the first instance – of a farmer, a businesswoman, a teacher, a retired couple – were all of white people and they were all impossibly attractive like superheroes or supermodels. We had to work through several iterations, instructing the programme to make the images more diverse, more 'ordinary', before we had something that was fit for purpose. Again, you need to know what questions to ask and how to prompt the programme. In the end, the images were appropriate and relatable, and we used them successfully to help a group of stakeholders think through issues related to climate change.

13.2 How AI can help build socially successful projects

So, at the time of writing, we feel that there is scope to use AI to support the development of socially successful projects, if handled with care.

To help us think through the relationship between AI, stakeholder engagement, and social value, we called on the expertise of Professor Toby Dubois-Heys, Professor of Digital Media in the School of Digital Arts at Manchester Metropolitan University (Dubois-Heys, 2025). He has worked in and around the field of AI for many years.

Through our conversation with Professor Dubois-Heys, it became clear that AI isn't going away. It is not something that we can choose not to use. It is already too ingrained in our daily life and we can't stop it… so we have to embrace it.

We have identified two ways to use AI to help enhance stakeholder engagement and social value processes on your project.

1 A Research tool

Early in the project, it can be beneficial to use an AI platform to support your research into stakeholders and social value. Professor Dubois-Heys explained that in principle, this offers a wider range of ideas and sources than consulting at an individual level, or even a few people such as locally knowledgeable colleagues (Dubois-Heys, 2025). The benefit of an AI model is that it can gather information from a larger number of sources, in a shorter space of time than we as a project team could ever hope to do. Some possible generic prompt questions to use to gain that first initial draft include:

- Who are the main stakeholders in the area who might be affected by the project?
- What are their social values?
- How should we approach engagement with these stakeholders?
- What are the engagement risks? Which of these risks most affects the social value of stakeholders?
- What might be the most effective engagement methods to use?

These are the types of questions that you would use in developing an engagement strategy. Using AI early in the project may help if you have little or no budget at this stage. However, this also opens up a risk, as in our view these resources should only be used as a support and prompt – offering a reasonable first draft – for the development of a plan which needs to involve your project team and the stakeholders themselves.

You can add additional prompts and revisit your research at later stages in the project. Prompt design is a key skill, and it is helpful to test and learn about how to achieve the best results (Amatriain, 2024). We find it particularly helpful to prompt for the sources of information and where possible go direct to that source.

You might want to go deeper by asking additional questions of the AI model such as:

- How might the community's social value help us improve the project in ways we have not thought of?
- How should this thinking affect how we organise our work as a project team?
- Who might be the 'forgotten stakeholders' for this project?

Further down the track you might ask, for instance, why stakeholders are not engaging and what could be done to encourage them. Or you could ask the AI model to explore how social value might evolve over time. Again, Professor Dubois-Heys explained that if you are using scenarios (see Chapter 8) AI could help develop the details of half a dozen scenarios much more quickly than a group of people could (Dubois-Heys, 2025). However, although you gain in time and breadth, you miss the experience, for instance, of a group of people working on scenarios together which is part of how you develop a functioning team or partnership.

Professor Dubois-Heys explained that AI has the power to put billions of data points together and gather all this data which we as a project team or humans alone just cannot do. It's this level of data that can significantly enhance your project, providing you with baseline evidence on which to start forming the relationships and having conversations with your stakeholders (Dubois-Heys, 2025).

2 *Framing communications*

The second use we identified was that AI platforms can be helpful in drafting communications to stakeholders.

Ask for a non-technical but authoritative piece of communication that will address the issues you have identified in your work to date, and you will receive a draft within a few seconds. Equally you can feed in data to produce a draft report to your specifications, a leaflet to distribute to residents, or the outline and text for a slide pack. You can refine your draft through multiple variations using prompt questions.

As we discussed above, you can also use AI to produce useful visual images. This can save time and effort.

We emphasise that these should be treated as drafts to be checked and reviewed by you, your team, and ideally members of your target audience. Look out for issues like diversity and accessibility in both text and visual imagery. Double check any references. The final versions of these documents should be yours and you need to be comfortable with your ownership and accountability for what is in them.

Another way AI can frame conversations is through actually engaging with stakeholders. An example is the engagement platform Hello Lamp Post. This AI-driven technology can share information with the public or provide an interactive touch point about your project (Hello Lamp Post, 2025a). This is usually done through QR codes positioned in locations on sites that are easily accessible to stakeholders and relevant to the project. The platform can share information or provide a two-way conversation with stakeholders who just scan the QR code with their mobile phones.

An example of this is an interactive art installation in downtown Iowa, USA. The artist Amanda Parer created 23-foot-tall, illuminated installations, including bunnies. Hello Lamp Post was used to tell the story of the bunnies with an underlying message of environmental conservation. Visitors to the installation scanned a QR code and were able to extract information and follow the story on their mobile phones. They were also able to provide feedback which allowed the client, Iowa City Downtown District, to gain visitor insights to use for future events (Hello Lamp Post, 2025b).

13.3 Two futures for AI in socially successful projects

No one knows how AI is going to shape our lives in the future, or how it will shape stakeholder engagement or social value practices, but in our modest and non-expert opinion we can see two potential scenarios:

1. AI could be used to handle the majority of engagement-related activities, saving on time and budget. It could provide a convenient shortcut to understanding the key issues for stakeholders and developing a plan. However, it would miss some of the core features of the approach we have outlined here, notably building relationships and tuning in to the tacit and emotional dimension of social value. Like any partial or incomplete engagement plan, it runs the risk of pushback from stakeholders and misses the opportunity to challenge your thinking and build genuine collaboration.

Or…

2. AI could provide support for the 'grunt work' on a project, freeing up time and resources for the really important stuff – listening, building relationships, understanding what is specific to your project at this particular moment, tailoring communication to an audience you know and understand. With support from AI, team members can focus on the more rewarding, intricate aspects of the work where human, rather than artificial intelligence, comes into its own. This type of work across sectors and projects could then inform the knowledge base we can draw upon over time. We believe that this is a potentially strong recipe for creating socially successful projects in the future.

Key takeaways

No one knows how AI is going to evolve and shape our project processes and practices. However, at present there is the potential to use AI to gather data from a wide range of sources very quickly to provide baseline evidence for your projects.

It is worth exploring engagement tools using AI.

Embracing AI as part of a wider engagement approach can help save time and enable a focus on what matters to the team and stakeholders.

References

Adams, D. (2020) *The Hitchhiker's Guide to the Galaxy*. Pan: London.

Amatriain, X. (2024) *Prompt design and engineering: Introduction and advanced methods*. [Online] Available at: https://arxiv.org/html/2401.14423v3. Accessed 25/06/2025.

Dubois-Heys, T. (2025) 'Book Interview'. Interview by Sarah Fitton [Microsoft teams] Unpublished, 12/6/2025.

Hello Lamp Post (2025a) *Powering AI digital assistants for any place* [Online] Available at: https://www.hlp.city/en-gb/about. Accessed 14/07/2025.

Hello Lamp Post (2025b) *Interactive art installation* [Online] Available at: https://www.hlp.city/en-gb/case-studies/interactive-art-installation. Accessed 14/07/2025.

Nonaka, I. (1994) A Dynamic theory of organizational knowledge creation. *Organization Science* 5:1, pp. 14–37. https://doi.org/10.1287/orsc.5.1.14

14 This isn't the end, this is the beginning

You cannot escape the responsibility of tomorrow by evading it today.
(Attributed to Abraham Lincoln)

14.1 Ambridge flood risk project 2.0

14.1.1 Year 1

14.1.1 January

Emily was surprised to find herself discussing climate change scenarios with the church warden... but quite a lot about the past month had been surprising.

She had been sitting at her desk working up the key points in the business case for a new flood scheme when her boss Gloria called her into the office and explained that the project needed to be socially successful.

'Socially successful?' she asked. 'What's that?'

Gloria explained her thinking about the project, which was to involve the community from the beginning, understand what mattered to them, take them on the journey, and even have them really influence the project design.

The first step was to work out who they meant by community.

She did some analysis, asked around and invited a group of local people including business owners who not only told her what they thought but suggested other people to talk to.

One of their suggestions was the church on the main shopping street which they said was a real hub for the community. Hence the chat with the churchwarden.

She also found herself questioning her assumptions. These early conversations had raised a lot of issues around heritage, the local economy, and community safety that she hadn't considered before. This was quite a different way of working to what she was normally used to, but she liked it.

On the other hand, she felt they were getting further from the answers as the questions became more complex. She couldn't find any guidance to show her where to go next or what to do.

DOI: 10.1201/9781003659877-19

14.1.1.2 March

Emily felt nervous at the meeting of the Project Board because she was presenting a paper based not only on her technical assessment, but on the set of discussions she'd been having.

The paper was titled 'What matters to people in Ambridge?' and the summary was: A lot of different things.

The most important point she wanted to get across was that flooding was seen as important, but it was by no means the most important issue. If their work cut across people's priorities around health and leisure, the cost of living, children's play spaces, and the complex heritage of this former mining town, they were unlikely to be supported.

The paper set out a detailed engagement plan, explaining that the plan had to be flexible and responsive to stakeholder views. This engagement plan had used a number of different stakeholder mapping and segmentation tools to create and was done earlier than usual. The team found they were referring to it more than they usually do.

Thankfully the Project Board listened to her paper respectfully and the discussion was really buzzy.

Gloria even suggested that they invite a couple of community representatives to join the Project Board. This seemed a bridge too far for some, but in the end, they agreed to consider this and asked Emily to find a way of recruiting some representative people.

14.1.1.3 June

George, the engineering consultant, was nervous about presenting first thoughts to the combined drop in/discussion sessions they were running in different community locations. 'The problem is', he said, 'I just won't have all the answers'.

Emily was really glad that they'd brought in an independent facilitator to help with these events. Nusret helped them think through what to communicate and how to handle questions and objections. She said 'everyone should have a say but that doesn't mean they can beat you up. I'll make sure that the discussion stays on track'. And she did.

Towards the end of one of the sessions held in a converted dance hall in the old theatre district, Emily saw George in deep conversation with some local residents. Afterwards he explained that they lived in one of the high-rise buildings on the edge of the area. They had described how run off from concrete paths between the high-rise blocks had worsened flooding back in the 1980s. 'I never knew there were eight high rises on the edge of town', said George, 'We need to go and talk to the people there'.

14.1.1.4 October

George had developed three options for the flood scheme. Before firming them up he ran them past the group he called the 'brains trust' – the churchwarden, the tenant group from the high-rise blocks and the local heritage champions. He was beginning to enjoy this aspect of the work, and had been on a course to learn how to have effective conversations.

One of the options involved new landscaping around the high-rise blocks. 'I'd never have come up with that through the technical analysis alone' he said. Emily and George presented the options to the Project Board. The community representatives congratulated them on the engagement so far. It seemed too good to be true...

14.1.2 Year 2

14.1.2.1 January

It was too good to be true. A series of workshops presenting the options for the flood scheme turned out to be quite stormy. Emily felt that only now was it 'getting real' for the community and they were particularly concerned about how long it would take to build the scheme, and the disruption during that time. Nusret's facilitation had really helped. None of the meetings degenerated, and Emily and George had kept their cool, but they realised they were missing something. They sat down over coffee and cake with the 'brains trust' which had now expanded to include members of the local mother and baby group and a couple of local shop owners who were very worried about the impact on trade of building works going on. They were all pretty vocal and Emily and George sat and listened.

'What could people live with, do you think? We can't avoid disruption with any of the options. But what would make it bearable?' Emily asked.

Gradually they worked out the shape of a plan for the construction phase including limits on the times of work, and an onsite contact person to handle any difficulties.

It also dawned on George that the community were not concerned about the options themselves, their attention had switched to the construction issues. The preferred option, including the high-rise landscaping, looked good to go.

The 'brains trust' agreed to work their networks and make clear that the team were still listening. 'We've known you a year now. We trust you. Just don't mess it up'.

14.1.2.2 April

Emily was eating crisps with one hand and writing with the other as she listened to teenagers living in the high rises talk about what they would like to see on their estate. A skate park seemed like the favourite idea, and the group were sketching out the ramps, bowls and ledges they wanted to see while Emily thought about how this could be included in the landscaping around the high rises.

The contractors had joined their meetings. They were now planning the actual build and the dialogue with the community was a bit overwhelming, but in a good way. Everyone seemed to have different ideas about what mattered: timing of works so as not to disturb people, keeping access open during construction, safety and in particular making sure that children were safe.

Mo from the contractor's team would be the key contact point during works and he was taking the time to get to know people. He handed out his mobile number so that people could get in touch during the works, if they had any issues.

14.1.2.3 June

Not exactly a ground-breaking ceremony, but Emily was up early and headed down to the site to see the official start of the works. Gloria brought coffee for her and Mo, and they went over the plans one last time.

As they were leaving the site, Emily said to Gloria, 'So is this now officially a socially successful project?' Gloria replied, 'We'll know in five years' time. Which reminds me, we need to look at the monitoring and evaluation plan'. But due to careful planning and accounting for the involvement of the community in the project process and the resources for that, Emily's project had come in on budget and just a little over schedule (due to some pesky newts).

14.2 Reflection

There is no traditional conclusion to this book.

We can't conclude this book with a nice, neat answer because we don't have a nice, neat problem. We have a messy complex situation, one that is being made worse by climate change and the associated impacts.

Socially successful projects aren't a nice to have, they are a necessity if we want to fight climate change, achieve the Sustainable Development Goals, create resilient communities, provide a better quality of life for those using the built environment, and give hope to future generations.

In this book we have made the case for giving equal consideration to the social, financial, and technical aspects of projects. This means engaging with the complex, messy, human parts that can't be quantified easily, and talking and listening to people. The difference this can make can be seen in The Ambridge flood risk project in Chapter 1 compared to the Ambridge flood risk project 2.0 you have just read.

We have focused on two mechanisms for achieving this: social value and stakeholder engagement. Both concepts are complex and contested in their own ways.

Social value is a relatively new concept within the built environment and consequently it is still finding its identity. Socially successful projects depend on identifying, creating, and maximising social values. Because it's new, the concept is open to interpretation, there is no clear industry-wide acceptance of a methodology, and its subjective and qualitative nature means it often goes against the grain of project processes and procedures.

Traditional stakeholder engagement practices are not fit for purpose anymore. They are project focused, compromised by power dynamics, and occur too late to provide meaningful insights that can shape a project. Engagement also tends to be the Cinderella element of projects and needs to be brought to the centre stage.

We advocate an approach to project design that builds on, and goes beyond, traditional engagement practices. Engagement should be early, authentic, and collaborative. It is about listening and hearing, not about announcing and responding. This enables social values to be identified early in the project so that they shape its design as well as build trust and lasting relationships with stakeholders.

The relationship between social value and stakeholder engagement is complex. A socially successful project can only be created with both. High-quality engagement allows social values to surface. These can then shape the project on equal terms with technical and financial considerations.

14.3 Principles for socially successful projects

To conclude this book, here is our high-level overview of the principles that guide socially successful projects. This is not a linear process, but if you apply these principles in your practice and on your projects, you are moving towards creating a socially successful project:

Principle 1: Built environment projects have a long lifespan, affecting this and future generations. Take time to understand what matters to the people who live, work, and play in a place.

Principle 2: Project teams need to focus on the social dimension of projects, especially social value and stakeholder engagement. If not, they risk opposition, delays, budget overruns, and reputational damage.

Principle 3: Focusing on the social dimension can be a challenge to current practice as it requires project managers and teams to state engagement objectives and listen to stakeholders early in projects, valuing the social dimension on equal terms with the technical and financial dimensions. Ingrain a social ethos in the project from the start. Champion a change to normal practice and routines.

Principle 4: Stakeholder engagement is not just a matter of process. It enables projects to grapple with complexity, the cultural, emotional, and practical concerns of stakeholders and communities. This needs to be realised and appreciated by project teams and different disciplines.

Principle 5: Thinking about social values helps us understand the wide range of benefits built environment projects can bring and enables us to tune in to what matters to stakeholders. Early on in your project, spend time listening to stakeholders to build relationships. Demonstrate and earn trust, be open and honest, and manage expectations.

Principle 6: Think about how to create safe spaces and methods for inclusive engagement and ensure that adequate time and resources are allocated to this.

Principle 7: Involve stakeholders in analysing the context of your project and scenarios you might face. This will help you to build greater resilience into your project planning.

Principle 8: Stakeholder mapping is a flexible, responsive, evolving activity. It can't, and shouldn't, be completed and left. Embrace the messiness and be open to the valuable insights a range of stakeholders can provide to the project.

Principle 9: Involve stakeholders in options development and appraisal. Include social criteria when appraising options.

Principle 10: Involve stakeholders in the monitoring and evaluation process. Qualitative data is as important as quantitative. Begin early in the project and ideally continue learning beyond the project end.

In our experience this shift of focus is challenging but worthwhile. Some of the case studies we have shared show what is possible.

We hope that this book has sparked your interest or provided the evidence you needed to follow these methods. Maybe you are thinking of being part of the socially successful project movement, for which we thank you.

Let's start creating socially successful projects, one project at a time, and before we know it, this way will be the norm, not the disruptor.

Index

For Product Safety Concerns and Information please contact our EU
representative GPSR@taylorandfrancis.com
Taylor & Francis Verlag GmbH, Kaufingerstraße 24, 80331 München, Germany

www.ingramcontent.com/pod-product-compliance
Ingram Content Group UK Ltd.
Pitfield, Milton Keynes, MK11 3LW, UK
UKHW020002200926
475702UK00015B/2436

9781041111955